菲迪克（FIDIC）文献译丛

客户/咨询工程师（单位）服务协议书范本

（原书2017年版）

Client/Consultant
Model Services Agreement

国际咨询工程师联合会 编

唐 萍 张瑞杰 等译

（正式使用发生争执时，以英文原版为准）

机械工业出版社

2017年版《客户/咨询工程师（单位）服务协议书范本》（白皮书）的条款由国际咨询工程师联合会（FIDIC，菲迪克）编写，该协议书适用于由雇主主导的设计团队以及由承包商主导的设计团队的设计和施工委托，一般用于投资前和可行性研究、详细设计以及施工管理和项目管理。该协议书适用于国际项目，但同样适用于国内项目。菲迪克（FIDIC）认为英文版是供翻译使用的正式的和权威性的文本。

2017年版白皮书强化了咨询工程师（单位）所承担的谨慎义务和职责。主要内容包括：
- 起草咨询协议书方面全球最新的实践；
- 客户与咨询工程师（单位）之间的公平风险平衡；
- 专业人员在适当技能、谨慎和适合用途方面的义务；
- 可用保险。

2017年版白皮书强化了咨询工程师（单位）对合理技能和谨慎的义务，咨询工程师应具有为同等规模和复杂性的项目提供服务的经验，在不扩展该义务的前提下，服务必须满足协议书规定的功能和目的。该义务不会使咨询工程师（单位）对因不可预见或无法控制的事件造成的有缺陷或不足的服务承担责任，因此，该义务应由任何职业责任保险保单进行充分保险。这既有利于客户又有利于咨询工程师（单位），并且代表了双方之间适当的风险平衡。

版权所有。未经出版者事先书面许可，对本出版物的任何部分不得以任何方式或途径复制或传播，包括但不限于复印、录制、录音，或通过任何数据库、信息或可检索的系统。

本书封面贴有机械工业出版社和国际咨询工程师联合会（FIDIC，菲迪克）双方的防伪标签，无标签或标签不全者不得使用和销售。

北京市版权局著作权合同登记 图字：01-2020-1319

图书在版编目（CIP）数据

客户/咨询工程师（单位）服务协议书范本：原书2017年版/瑞士国际咨询工程师联合会编；唐萍等译. —北京：机械工业出版社，2021.3

（菲迪克（FIDIC）文献译丛）

书名原文：Client/Consultant Model Services Agreement

ISBN 978-7-111-67684-3

Ⅰ. ①客… Ⅱ. ①瑞… ②唐… Ⅲ. ①建筑工程-咨询服务-协议-范本 Ⅳ. ①TU712

中国版本图书馆CIP数据核字（2021）第038776号

机械工业出版社（北京市百万庄大街22号　邮政编码100037）
策划编辑：何文军　责任编辑：何文军　李宣敏
责任校对：陈　越　封面设计：张　静
责任印制：李　昂
北京铭成印刷有限公司印刷
2021年5月第1版第1次印刷
210mm×297mm・6印张・275千字
标准书号：ISBN 978-7-111-67684-3
定价：99.00元

电话服务　　　　　　网络服务
客服电话：010-88361066　机　工　官　网：www.cmpbook.com
　　　　　010-88379833　机　工　官　博：weibo.com/cmp1952
　　　　　010-68326294　金　　书　　网：www.golden-book.com
封底无防伪标均为盗版　机工教育服务网：www.cmpedu.com

译者的话

本书是由国际咨询工程师联合会（FIDIC，菲迪克）编写，于 2017 年出版的第 5 版《客户 / 咨询工程师（单位）服务协议书范本》（白皮书）的中英文对照本，代表客户与其咨询工程师（单位）之间委托合同的基本格式，旨在满足典型委托合同的最低要求。专用条件中可能需要附加或经修订的条款，以解决各方之间的特定项目和商业问题。2017 年版白皮书强化了咨询工程师（单位）所承担的谨慎义务和职责。在咨询协议书起草方面参考了全球最新实践，客户与咨询工程师（单位）之间的风险分配更加平衡，更加强调专业人员在适当技能、谨慎和适合用途方面的义务，增加了可用保险的相关条款。

希望本书的出版，对我国广大从事工程咨询（设计）、投资、金融和项目管理的部门和组织，各类项目业主，建筑施工监理企业，工程承包企业，环保企业，会计 / 律师事务所，保险公司以及有关高等院校等人员在学习和运用菲迪克合同条件时，能有效地解决在国际、国内工程咨询和工程承包活动中的合同管理问题，更好地开拓国内外工程咨询和工程承包市场，对提高工程建设的社会效益和投资效益，建立和完善工程项目总承包制度，促进我国工程建设管理体制与国际惯例接轨，推动我国工程建设管理模式和体制全面深化改革有所帮助。

翻译过程中，我们虽然尽力想使译文准确通顺，完整地传达原文的内容，汉语表达规范易懂，但限于专业知识与语言水平，译文中可能出现不妥乃至错误之处，敬请读者指正。本书以中英文对照方式编排，以便用户核对中译文，从而更准确地理解白皮书。

本书由唐萍、张瑞杰、贾志成、史骏、邓冰茹、张辰旭、郭文涛、李莉萍、莫伟平、郑海燕、秦春燕、曾家平、张荣芹等翻译，唐萍、张瑞杰、贾志成校译，唐萍、张瑞杰、邓冰茹、张辰旭审校。

译者

FIDIC is the international federation of national Member Associations of consulting engineers.

FIDIC was founded in 1913 by three national associations of consulting engineers within Europe. The objectives of forming the Federation were to promote in common the professional interests of the Member Associations and to disseminate information of interest to their members. Today, FIDIC membership covers some 90 countries from all parts of the globe encompassing most of the private practice consulting engineers.

FIDIC is charged with promoting and implementing the consulting engineering industry's strategic goals on behalf of Member Associations. Its strategic objectives are to: represent world-wide the majority of firms providing technology-based intellectual services for the built and natural environment; assist members with issues relating to business practice; define and actively promote conformance to a code of ethics; enhance the image of consulting engineers as leaders and wealth creators in society; promote the commitment to environmental sustainability; support and promote young professionals as future leaders.

FIDIC arranges seminars, conferences and other events in the furtherance of its goals: maintenance of high ethical and professional standards; exchange of views and information; discussion of problems of mutual concern among Member Associations and representatives of the international financial institutions; and development of the consulting engineering industry in developing countries.

FIDIC members endorse FIDIC's statutes and policy statements and comply with FIDIC's Code of Ethics which calls for professional competence, impartial advice and open and fair competition.

FIDIC, in the furtherance of its goals, publishes international standard forms of contracts for works (Short Form, Construction, Plant and Design Build, EPC/Turnkey, Design, Build and Operate) and agreements (for clients, consultants, sub-consultants, joint ventures, and representatives), together with related materials such as standard pre-qualification forms.

FIDIC also publishes business practice documents such as policy statements, position papers, guides, guidelines, training manuals, and training resource kits in the areas of management systems (quality management, risk management, integrity management, sustainability management) and business processes (consultant selection, quality based selection, tendering, procurement procedure, insurance, liability, technology transfer, capacity building, definition of services).

FIDIC organises an extensive programme of seminars, conferences, capacity building workshops, and training courses.

FIDIC aims to maintain high ethical and professional standards throughout the consulting engineering industry through the exchange of views and information, with discussion of problems of mutual concern among Member Associations and representatives of the multilateral development banks and other international financial institutions.

FIDIC publications and details about training courses and conferences are available from the Secretariat in Geneva, Switzerland. Specific activities are detailed in an annual business plan and the FIDIC website, www.fidic.org, gives extensive background information.

菲迪克（FIDIC）是咨询工程师国家（地区）成员协会国际联合会。

菲迪克（FIDIC）是由欧洲三个国家的工程咨询协会于 1913 年成立的。组建联合会的目的是共同促进成员协会的职业利益，以及向其会员传播有益信息。今天，菲迪克（FIDIC）已有来自于全球各地 90 个国家（地区）的会员，包括大多数私人执业的咨询工程师。

菲迪克（FIDIC）代表成员协会负责促进和实施工程咨询行业的战略目标。其战略目标是：代表全世界为建设和自然环境提供以技术为基础的智力服务的大多数公司；协助会员处理与业务实践相关的问题；制定并积极促进遵守职业道德规范；提升咨询工程师作为社会领导者和财富创造者的形象；促进对环境可持续性的承诺；支持和促进青年专业咨询工程师成为未来的领导者。

菲迪克（FIDIC）举办各类研讨会、会议及其他活动，以促进其目标：维护高水平的道德和职业标准；交流观点和信息；讨论成员协会和国际金融机构代表共同关心的问题；以及发展中国家工程咨询业的发展。

菲迪克（FIDIC）会员认可菲迪克（FIDIC）章程和政策声明，并遵守其职业道德规范要求的专业技能、公正的建议和公开公平的竞争。

菲迪克（FIDIC）为了实现其目标，发布了国际标准格式的工程合同（简明格式、施工、生产设备和设计-施工、EPC/交钥匙、设计、施工和运营）和协议书［针对客户、咨询工程师（单位）、分包咨询工程师、联营体和代表］，以及资格预审标准格式等相关资料。

菲迪克（FIDIC）还出版比如政策声明、行动报告、指南、指导方针、培训手册和管理体系领域的培训资料包（质量管理、风险管理、廉洁管理、可持续管理）以及业务流程［咨询工程师（单位）的选择、根据质量选择（咨询服务）、招标、采购程序、保险、责任、技术转让、实力建设、服务定义］的业务实践文件。

菲迪克（FIDIC）组织研讨会、会议、实力建设研讨会和培训课程等各类活动。

菲迪克（FIDIC）致力于通过交换观点和信息，在成员协会和多边开发银行及其他国际金融机构代表之间讨论共同关心的问题，在整个工程咨询行业维护高水平的道德和职业标准。

菲迪克（FIDIC）出版物以及培训课程和会议的详细信息，可以从设在瑞士日内瓦的菲迪克（FIDIC）秘书处得到。具体活动详见年度业务计划和菲迪克（FIDIC）网站，www.fidic.org，该网站提供了大量的背景信息。

COPYRIGHT

© FIDIC 2021 All rights reserved.

The Copyright owner of this document is the International Federation of Consulting Engineers - FIDIC.

Translation from English to Chinese has been performed by China Machine Press with FIDIC's permission.

The lawful purchaser of this document has the right to make a single copy of the duly purchased document for his or her personal and private use. No part of this publication may be shared reproduced, distributed, translated, adapted, stored in a retrieval system, or communicated, in any form or by any means, mechanical, electronic, magnetic, photocopying, recording or otherwise, without prior written permission from FIDIC. To request such permission, please contact FIDIC, Case 311, CH-1215 Geneva 15, Switzerland; fax +41 22 799 49 01, e-mail: fidic@fidic.org. Electronic copies can be obtained from FIDIC at www.fidic.org/bookshop.

FIDIC considers the official and authentic text to be the version in the English language and assumes no liability whatsoever for the completeness, correctness, adequacy or otherwise of the translation into Chinese for any use to which this document may be put.

Disclaimer

While FIDIC aims to ensure that its publications represent the best in business practice, the Federation accepts or assumes no liability or responsibility for any events or the consequences thereof that derive from the use of its publications, including their translations. FIDIC publications are provided "as is" without warranty of any kind, either express or implied, including, without limitation, warranties of merchantability, fitness for a particular purpose and non-infringement. FIDIC publications are not exhaustive and are only intended to provide general guidance. They should not be relied upon in a specific situation or issue. Expert legal advice should be obtained whenever appropriate, and particularly before entering or terminating a contract.

版权

©FIDIC 2021 版权所有。

本文件的版权所有人为国际咨询工程师联合会——菲迪克（FIDIC）。

经菲迪克（FIDIC）许可，中国机械工业出版社完成了英文版的中文翻译。

本文件的合法购买者有权将正式购买的文件制作副本，供其个人和私人目的使用。未经菲迪克(FIDIC) 事先书面许可，不得将本出版物的任何部分分享复制、分发、翻译、改编、存储在检索系统中，或以任何格式或通过任何方式以机械、电子、磁性、影印、记录或其他方式传送。如有意获得此类许可，请联系菲迪克（FIDIC），地址：Case 311, CH-1215 Geneva 15, Switzerland；传真：+41 22 799 49 01，电子邮件：fidic@fidic.org。电子版本可从菲迪克（FIDIC）获取，网址为www.fidic.org/bookshop。

菲迪克（FIDIC）认为正式的和权威性的文本为英语版本，并对本文件的任何用途的中文翻译文本的完整性、正确性、充分性或其他方面不承担任何责任。

免责声明

尽管菲迪克（FIDIC）的目标是确保其出版物代表最佳业务实践，但对于使用其出版物及其翻译文本而引起的任何事件或后果，联合会（菲迪克，FIDIC）不承担任何责任或义务。菲迪克（FIDIC）出版物按"原样"提供，没有任何明示或暗示的保证，包括对可销售性、特定用途的适用性和非侵权性的无限保证。菲迪克（FIDIC）出版物并非详尽无遗，仅提供一般性指导。在特定的情况或问题上不应依赖它们。除此之外，适当时，尤其在签订或终止合同前，应获得专家法律建议。

ACKNOWLEDGEMENTS

The Federation Internationale des Ingenieurs-Conseils (FIDIC) extends special thanks for the preparation of this Fifth Edition 2017 of the Client/Consultant Model Services Agreement to the members of its Task Group: Mike Roberts (Task Group Leader and Principal Drafter), Consultant, Whiskers LPP, UK; Vincent Leloup, Managing Partner, Exequatur, France; Juan Carlos Moncada Bueno, TYPSA, Spain; Ulla Sassarsson, FRI, Denmark; Jeroen van Gessel, Witteveen&Bos, Netherlands.

The preparation was carried out under the general direction of the FIDIC Contracts Committee which comprised Philip Jenkinson (past chair), Christoph Theune (past member), Kaj Möller, Siobhan Fahey, José Amorim Faria (past EFCA liaison), Mike Roberts, Des Barry, Vincent Leloup, William Howard, PawelZejer and Zoltán Záhonyi.

Draft documents were reviewed and valuable comments received from the following people and organisations: Mahmoud Abu Hussein, Dolphin Energy, UAE; Syntec Ingénierie, France; Benoît Chassatte, French Development Agency (AFD); European International Contractors; Christopher Wright, Christopher Wright and Co. LLP, England; Ronnie Thompson, AECOM, Hong Kong; Ulrik Bang-Olsen, Bang-Olsen & Partners Law Firm, Denmark; William Godwin, 3 Hare Court, England. FIDIC would like to thank the above persons and organisations for their contributions to the Fifth Edition.

FIDIC would also like to thank Charles Nairac, White and Case, Paris for performing a thorough legal review of the final draft document, and Margaret Walker, MWCAM Ltd, for her editing services.

The ultimate decision on the form and content of the publication Client/Consultant Model Services Agreement rests with FIDIC.

致谢

国际咨询工程师联合会（FIDIC，菲迪克）向编写 2017 年第 5 版《客户/咨询工程师（单位）服务协议书范本》工作组的以下成员特致谢意：英国 Whiskers LPP 公司的咨询工程师 Mike Roberts（工作组组长兼主要起草人）、法国 Exequatur 公司的执行合伙人 Vincent Leloup、西班牙 TYPSA 公司的 Juan Carlos Moncada Bueno、丹麦 FRI 公司的 Ulla Sassarsson、荷兰 Witteveen&Bos 公司的 Jeroen van Gessel。

编写工作在菲迪克（FIDIC）合同委员会的总体指导下开展，包括 Philip Jenkinson（前董事长）、Christoph Theune（前委员）、Kaj Möller、Siobhan Fahey、José Amorim Faria（前 EFCA 联络人）、Mike Roberts、Des Barry、Vincent Leloup、William Howard、PawelZejer 以及 Zoltán Záhonyi。

下列人员和组织对初稿进行了审阅并提出宝贵意见：阿拉伯联合酋长国 Dolphin Energy 公司的 Mahmoud Abu Hussein；法国工程咨询协会；法国开发署（AFD）Benoît Chassatte；欧洲国际承包商协会；英国 Christopher Wright and Co LLP 公司的 Christopher Wright；AECOM 公司的 Ronnie Thompson；丹麦 Bang-Olsen & Partners 律师事务所的 Ulrik Bang-Olsen；英国 3 Hare Court 商会的 William Godwin。菲迪克（FIDIC）向上述人员和组织对第 5 版的贡献表示非常感谢。

菲迪克（FIDIC）还要向巴黎 White and Case 公司的 Charles Nairac 对文件终稿的法律审核，以及 MWCAM Ltd 公司的 Margaret Walker 提供的编辑服务表示感谢。

本客户/咨询工程师（单位）服务协议书范本格式和内容的最终决定由菲迪克（FIDIC）负责。

COMPLETION OF THE AGREEMENT

This Client/Consultant Model Services Agreement represents the basic form of a Contract of Appointment between a Client and its Consultant. It is intended to cover the minimum requirements of a typical appointment contract. Additional or amended clauses may be required in the Particular Conditions to address particular project and commercial issues between the parties.

Where other material is to be incorporated into a Contract of Appointment, care must be taken to ensure consistency both in the use of terminology and the allocation of duties and obligations. The parties to the contract may wish to take independent legal advice in connection with the preparation of this agreement. Independent legal advice may also help the parties understand their legal liabilities, duties and obligations arising under the Model Services Agreement.

Neither FIDIC nor any committee or individual connected with FIDIC can be held liable for project or commercial losses suffered as a result of adopting the Client/Consultant Model Services Agreement as the basis of a contractual arrangement.

协议书的编写

本**客户/咨询工程师（单位）服务协议书范本**代表**客户**与其**咨询工程师（单位）**之间**委托合同**的基本格式。它旨在满足典型委托合同的最低要求。**专用条件**中可能需要附加或经修订的条款，以解决各方之间的特定项目和商业问题。

如果将其他资料编入**委托合同**，则必须注意确保术语使用以及任务和义务分配的一致性。合同各方可能希望征询与本协议书的编制相关的独立的法律意见。独立的法律咨询还可帮助各方了解其在**服务协议书范本**中规定的法律责任、任务和义务。

菲迪克（FIDIC）或与其相关的任何委员会或个人均不对因采用**客户/咨询工程师（单位）服务协议书范本**作为合同安排的依据而遭受的项目或商业损失承担责任。

FOREWORD

The terms of the Client/Consultant Model Services Agreement (the White Book) have been prepared by the Federation Internationale des Ingenieurs-Conseils (FIDIC). The Agreement is suitable for general use for the purposes of pre-investment and feasibility studies, detail design, and administration of construction and project management, both for Employer-led design teams and for Contractor-led design teams on design and build commissions. The Agreement is suitable for international projects but can equally be used on domestic projects. The version in English is considered by FIDIC as the official and authentic text for the purpose of translation.

In the preparation it was recognised that whilst there are numerous clauses which will be generally applicable there are some provisions which must necessarily vary to take account of the circumstances and locality in which the Services are to be performed. The clauses of general application are under General Conditions. They are intended for incorporation as printed in the documents comprising the Agreement.

The General Conditions are linked with the Particular Conditions by the corresponding numbering of the clauses so that the General Conditions and Particular Conditions together comprise the conditions governing the rights and obligations of the parties.

The Particular Conditions must be specially drafted to suit each individual Agreement and type of Service. References to the General Conditions are in Part A whilst any amendments to the General Conditions or additional clauses are in Part B. These pages must be completed for incorporation into the Agreement.

FIDIC intends to publish an updated "White Book Guide" which will include comment on the clauses in the White Book and guidance on completion of the Appendices. Compilers of the White Book in the meantime can refer to various other FIDIC publications available in the Bookshop of FIDIC's website at www.fidic.org. Compilers of scope of work in Appendix 1 may wish to consult the FIDIC Definition of Services guide.

This Fifth Edition of the White Book has enhanced the duty of care obligations placed on the Consultant. The approach taken by the Task Group and FIDIC Contracts Committee (CC) on this matter is illustrated below:

Consultant's Obligations for the Services.

The Terms of Reference laid down by the FIDIC CC for updating the White Book required the Task Group to consider, amongst many other matters, the issue of the professional's duty of care to the client for the services rendered. In this endeavour the Task Group considered the following, without limitation:

- up to date practice worldwide in drafting consultancy agreements
- a fair balance of risk between the Client and the Consultant
- the professional's obligation in respect to due skill and care and fitness for purpose
- available insurance

前言

客户/咨询工程师（单位）服务协议书范本（**白皮书**）的条款已由**国际咨询工程师联合会**（FIDIC，菲迪克）编制。该**协议书**适用于由**雇主**主导的设计团队以及由**承包商**主导的设计团队的设计和施工委托，一般用于投资前和可行性研究、详细设计以及施工管理和项目管理。该**协议书**适用于国际项目，但同样适用于国内项目。菲迪克（FIDIC）认为英文版是供翻译使用的正式的和权威性的文本。

在编制工作中，人们认识到，尽管有很多条款普遍适用，但有些条款必须要考虑履行**服务**的环境和地点做必要的变更。**通用条件**中包括了一般适用的条款，旨在按照原样纳入组成该**协议书**的文件。

通用条件通过条款的相应编号与**专用条件**相衔接，以便**通用条件**和**专用条件**共同构成管理各方权利和义务的条件。

专用条件必须专门起草以便与每一独特**协议书**和**服务**类型相适应。**通用条件**的引用在 A 部分，而对**通用条件**或附加条款的任何修改均在 B 部分。必须填写这些页面才能将其纳入**协议书**。

菲迪克（FIDIC）拟出版修订的"**白皮书指南**"，其中将包括对**白皮书**条款的解释以及填写完成**附录**的指南。同时，**白皮书**的编写者可以参考菲迪克（FIDIC）网上书店（网址：www.fidic.org）的各种其他菲迪克（FIDIC）出版物。**附录 1** 中的服务范围的编写者可能希望参考菲迪克（FIDIC）**服务定义**指南。

第 5 版**白皮书**强化了**咨询工程师**（单位）所承担的谨慎义务的职责。工作组和菲迪克（FIDIC）合同委员会（CC）在此问题上采取的方法如下：

咨询工程师（单位）的服务义务。

菲迪克（FIDIC）**合同委员会**制订的修订**白皮书**的**工作大纲**要求**工作组**在众多其他事项中，考虑专业人员向客户提供服务的谨慎义务。在这项工作中，**工作组**考虑了以下内容，但不限于：

- 起草咨询协议书方面全球最新的实践；
- **客户**与**咨询工程师**（单位）之间的公平风险平衡；
- 专业人员在适当技能、谨慎和适合用途方面的义务；
- 可用保险。

The Task Group recognised considerable pressure from some parts of the industry to enhance the obligations placed on the Consultant to ensure that the professional services and deliverables would be fit for purpose. The problem confronting the Task Group is that there is no common understanding of due skill and care or fitness for purpose either between clients and consultants or between legal advisors in various jurisdictions notwithstanding the wide usage of such terms.

In looking at this problem the Task Group accepted that the Client is entitled to expect that the professional services will be completed correctly and that all the specific contractual requirements will be met, and that if they are not correct, or any requirement has not been fulfilled, then the Client should be entitled to appropriate redress against the Consultant. This approach assumes that the Consultant is fully experienced and competent in the delivery of the relevant services and accordingly an appropriate stipulation has been added to the new White Book.

In assessing a fair balance of risk between the Client and the Consultant the Task Group considered the position where the services were rendered defective or inadequate for reasons beyond the Consultant's control or knowledge.

It is well known in the industry that Consultant's cover their liability under contract by taking out professional indemnity insurance. The Task Group, and the FIDIC CC, is satisfied that professional indemnity insurance policies do not cover liability for defective or inadequate services without evidence of fault or breach on the part of the Consultant. Such insurance only covers liability where there is a failure on the part of the Consultant to use reasonable skill and care to be expected from an experienced consultant.

The Task Group also examined whether the normal obligation placed on Consultants to use reasonable skill and care in delivering the services was an industry standard. The Task Group examined over 20 standard forms of appointment for consultants worldwide and noted that none of these forms of appointment required professional services to be fit for purpose or imposed strict liability for defects. The standard, whether expressed or implied, was reasonable skill and care. Accordingly, the Task Group determined that it was not a fair or reasonable balance of risk to make the Consultant strictly liable for the outcome of the professional services in situations where there was no evidence of fault or breach on its behalf.

The Task Group noted that in certain civil law jurisdictions strict liability for defective services was imposed on the Consultant – in some jurisdictions relief was available for matters that were not within the Consultant's control whereas in others no relief was available. The Task Group acknowledged, that in a limited number of countries, provision of insurance to cover strict liability was a mandatory obligation placed on the insurance market, however this approach is not followed generally and cannot be taken to reflect the international position. The Task Group and FIDIC CC determined therefore that the appropriate standard of care to be imposed on a Consultant was that of reasonable skill and care to be expected from an experienced consultant.

The Fifth Edition has enhanced the Consultant's obligation to the reasonable skill and care to be expected from a consultant experienced in the provision of services for projects of similar size and complexity and further, but without extending this obligation, the services must satisfy the function and purpose described in the Agreement. This obligation will not make the Consultant liable for defective or inadequate services arising out of unforeseeable or uncontrollable events and therefore, the obligation should be fully insured under any professional indemnity insurance policy. This is to the benefit of both the Client and the Consultant and represents the correct balance of risk between the two parties.

工作组认识到行业的某些领域要求强化**咨询工程师（单位）**所承担义务的巨大压力，以确保专业服务和交付成果达到适合用途。**工作组**面临的问题是，尽管这些术语的使用范围很广，但在客户和咨询工程师（单位）之间，或在各个司法管辖区的法律顾问之间，对于适当的技能和谨慎或适合用途尚无共识。

在研究该问题时，**工作组**认为**客户**有权期望专业服务将正确完成，以及所有特定合同要求均能得到满足，如果不正确或任何一项要求未得到满足，则**客户**应有权要求**咨询工程师（单位）**进行适当赔偿。这一方法是假定**咨询工程师（单位）**在提供相关服务方面具有丰富的经验和能力，因此在新的**白皮书**中增加了适当的规定。

在评估**客户**与**咨询工程师（单位）**之间的公平风险平衡时，**工作组**考虑了因**咨询工程师（单位）**无法控制或不了解的原因而导致服务出现缺陷或不足的地方。

业内周知，**咨询工程师（单位）**通过购买职业责任保险来承保合同规定的责任。**工作组**和**菲迪克（FIDIC）合同委员会**感到满意的是，如果没有提供**咨询工程师（单位）**的过错或违约的证据，职业责任保险保单则不承保因服务缺陷或不足而产生的责任。此类保险仅承保**咨询工程师（单位）**未能使用所期望的经验丰富的咨询工程师的合理技能和谨慎的责任。

工作组还审查了**咨询工程师（单位）**在提供服务时承担的使用合理技能和谨慎的正常义务是否为行业标准。**工作组**审查了全球 20 多种委托咨询服务的标准格式，注意到这些委托格式均不要求专业服务满足适用目标或对缺陷施加严格责任。无论是明示的还是暗含的，标准都是合理的技能和谨慎。因此，**工作组**认为，在没有其过错或违约证据的情况下，使**咨询工程师（单位）**对专业服务的结果承担严格责任不是公平或合理的风险平衡。

工作组注意到，在某些大陆法系司法管辖区，缺陷服务的严格责任施加给了**咨询工程师（单位）**——在某些司法管辖区，对于不在**咨询工程师（单位）**控制范围内的事务可减免责任，而在其他司法管辖区，则无减免。**工作组**承认，在少数国家，提供保险承保严格责任是保险市场的一项强制性义务，但该做法未能普遍遵循，也未能采用这种方法来反映国际地位。因此，**工作组**和**菲迪克（FIDIC）合同委员会**认为，**咨询工程师（单位）**应承担的适当谨慎标准是经验丰富的咨询工程师（单位）应具备的合理技能和谨慎。

第 5 版强化了**咨询工程师（单位）**对合理技能和谨慎的义务，咨询工程师应具有为同等规模和复杂性的项目提供服务的经验，在不扩展该义务的前提下，服务必须满足**协议书**规定的功能和目的。该义务不会使**咨询工程师（单位）**对因不可预见或无法控制的事件造成的有缺陷或不足的服务承担责任，因此，该义务应由任何职业责任保险保单进行充分保险。这既有利于**客户**又有利于**咨询工程师（单位）**，并且代表了双方之间正确的风险平衡。

CONTENTS

Form of Agreement 2

Particular Conditions 4

Part A   References from Clauses in the General Conditions

Part B   Additional or Amended Clauses

Appendices

        1   Scope of Services

        2   Personnel, Equipment, Facilities and Services of Others to be Provided by the Client

        3   Remuneration and Payment

        4   Programme

        5   Rules for Adjudication

Client/Consultant Model Services Agreement

General Conditions 30

      CONTENTS

1	GENERAL PROVISIONS	30
2	THE CLIENT	44
3	THE CONSULTANT	48
4	COMMENCEMENT AND COMPLETION	52
5	VARIATIONS TO SERVICES	56
6	SUSPENSION OF SERVICES AND TERMINATION OF AGREEMENT	58
7	PAYMENT	64
8	LIABILITIES	68
9	INSURANCE	70
10	DISPUTES AND ARBITRATION	72

目录

| 协议书格式 | 3 |
| 专用条件 | 5 |

A 部分　　参照通用条件有关条款

B 部分　　附加或修订的条款

附录

  1　服务范围

  2　要由客户提供的人员、设备、设施和其他方的服务

  3　报酬和付款

  4　进度计划

  5　裁决规则

客户/咨询工程师（单位）服务协议书范本

通用条件 ··· 31

  目录

1	一般规定	31
2	客户	45
3	咨询工程师（单位）	49
4	开始和完成	53
5	服务变更	57
6	服务暂停和协议书终止	59
7	付款	65
8	责任	69
9	保险	71
10	争端和仲裁	73

Form of Agreement

Between [Name of Client] ....

of [Address of Client] ....

(hereinafter called "the Client")

and [Name of Consultant] ....

of [Address of Consultant] ....

(hereinafter called "the Consultant")

WHEREAS:

The Client desires that certain Services should be performed by the Consultant, namely:
[Brief description of Services]

and has accepted an offer/proposal by the Consultant for the performance of such Services.

THE CLIENT AND THE CONSULTANT AGREE AS FOLLOWS:

1. In the Agreement words and expressions shall have the same meanings as are respectively assigned to them in Clause 1.1 of the General Conditions of the Client/Consultant Model Services Agreement.

2. The following documents shall be deemed to form and be read and construed as part of the Agreement and shall be given the order of precedence as below:

   (a) This Form of Agreement;
   (b) The Client/Consultant Model Services Agreement;
       (i) Particular Conditions;
       (ii) General Conditions;
   (c) Appendices 1 to 5;
   (d) Any letter of acceptance by the Client incorporated into the Agreement under Sub-Clause 1.1.1; and
   (e) Any letter of offer/proposal by the Consultant incorporated into the Agreement under Sub-Clause 1.1.1.

3. In consideration of the payments to be made by the Client to the Consultant under the Agreement, the Consultant hereby agrees with the Client to perform the Services in conformity with the provisions of the Agreement.

4. The Client hereby agrees to pay the Consultant in consideration of the performance of the Services such amounts as may become payable under the provisions of the Agreement at the times and in the manner prescribed by the Agreement.

AUTHORISED SIGNATURE(S) OF CLIENT: AUTHORISED SIGNATURE(S) OF CONSULTANT:

Signature ....   Signature ....

Name ....   Name ....

Position ....   Position ....

Date ....   Date ....

协议书格式

　　　　　　由　　　　　［客户名称］　　...

　　　　　　　　　　　　［客户地址］　　...

（下称"客户"）

　　　　　　和　　　　　［咨询工程师（单位）名称］　...

　　　　　　　　　　　　［咨询工程师（单位）地址］　...

［下称"咨询工程师（单位）"］之间

鉴于：

客户欲请**咨询工程师（单位）**履行一定的**服务**，即：
［**服务**简述］

并已接受**咨询工程师（单位）**为履行上述**服务**所提出的报价/建议书。

客户和**咨询工程师（单位）**兹达成协议如下：

1　　本**协议书**中的词语和措辞应与**客户/咨询工程师（单位）服务协议书**范本通用条件的第1.1款中分别赋予它们的含义相同。

2　　下列文件应被视为构成本**协议书**的组成部分，应作为其一部分阅读和解释，并应按以下顺序排序：

　　（a）　本**协议书**格式；
　　（b）　**客户/咨询工程师（单位）服务协议书**范本；
　　　　　（i）专用条件；
　　　　　（ii）通用条件；
　　（c）　附录1～附录5；
　　（d）　根据第1.1.1项规定纳入**协议书**的**客户**的任何中标函；（以及）

　　（e）　根据第1.1.1项规定纳入**协议书**的**咨询工程师（单位）**的任何报价/建议书。

3　　鉴于**客户**将根据本**协议书**的规定给**咨询工程师（单位）**付款，**咨询工程师（单位）**在此同意遵照本**协议书**的规定向**客户**履行**服务**。

4　　鉴于**咨询工程师（单位）**履行的**服务**，**客户**在此同意按本**协议书**规定的时间和方式，向其支付本**协议书**规定的应付款项。

客户授权代表签字：　　　　　　　　　　　　　　**咨询工程师（单位）授权代表签字：**

签字　...............................　　　　　　　　　　签字　...............................

姓名　...............................　　　　　　　　　　姓名　...............................

职务　...............................　　　　　　　　　　职务　...............................

日期　...............................　　　　　　　　　　日期　...............................

Particular Conditions

Part A. References from Clauses in the General Conditions

1.1 Definitions

1.1.4	Client's Representative	[Name of Representative]
1.1.5	Commencement Date	[Number of days] days after Effective Date
1.1.8	Consultant's Representative	[Name of Representative]
1.1.9	Country	[Name of Country]
1.1.22	Project	[Name of Project]
1.1.24	Time for Completion	[Time in days]

1.3 Notices and other Communications

1.3.1(c) Communication — [System of electronic communication accepted]

1.3.1(d) Address for communications

Client's address: [Address]

Email: (only when e-mail is accepted as a valid system for electronic communications) [Email]

Facsimile number: [Number]

Consultant's address: [Address]

Email: (only when e-mail is accepted as a valid system for electronic communications) [Email]

Facsimile number: [Number]

1.4 Law and Language

1.4.1	Law governing Agreement	[Law]
1.4.2	Ruling language of Agreement	[Language]
1.4.3	Language for communications	[Language]

1.8 Confidentiality

1.8.3	Period for expiry of confidentiality	[Years if different than Two]

1.9 Publication

1.9.1	Publication restrictions	[State restrictions on publication, if any]

专用条件

A 部分　　　参照通用条件有关条款

1.1 定义

1.1.4　客户代表　　　　　　　　　　　　[代表姓名]

1.1.5　开始日期　　　　　　　　　　　　[天数] 生效日期后的天数

1.1.8　咨询工程师（单位）代表　　　　　[代表姓名]

1.1.9　工程所在国　　　　　　　　　　　[工程所在国名称]

1.1.22　项目　　　　　　　　　　　　　 [项目名称]

1.1.24　完成时间　　　　　　　　　　　 [按天算的时间]

1.3 通知和其他通信交流

1.3.1（c）　通信交流　　　　　　　　　 [接受的电子通信交流系统]

1.3.1（d）　通信交流地址

客户地址：　　　　　　　　　　　　　　 [地址]

邮件：（仅在电子邮件被接受为有效电子通信交流系统时）
　　　　　　　　　　　　　　　　　　　 [邮件]

传真号码：　　　　　　　　　　　　　　 [号码]

咨询工程师（单位）地址：　　　　　　　 [地址]

邮件：（仅在电子邮件被接受为有效电子通信交流系统时）
　　　　　　　　　　　　　　　　　　　 [邮件]

传真号码：　　　　　　　　　　　　　　 [号码]

1.4 法律和语言

1.4.1　管辖协议书的法律　　　　　　　　[法律]

1.4.2　协议书的主导语言　　　　　　　　[语言]

1.4.3　通信交流语言　　　　　　　　　　[语言]

1.8 保密

1.8.3　保密有效期限　　　　　　　　　　[如果不用于两年，则年数]

1.9 出版物

1.9.1　出版限制　　　　　　　　　　　　[述明出版限制，如果有]

3.9	**Construction Administration**	[Included in Services/Not included in Services]
7.4	**Third Party Charges on Consultant**	[Exemption Applies/Exemption does not Apply]
8.2	**Duration of Liability**	
8.2.1	Period of Liability	[Period]
8.3	**Limit of Liability**	
8.3.1	Limit of Liability	[Amount]
9	**Insurance**	
9.1.1	Insurances to be taken out by Consultant	
	Professional Indemnity Insurance	[Amount]
	Public Liability Insurance	[Amount]
10	**Disputes and Arbitration**	
10.4.1	Arbitration rules	[International Chamber of Commerce (or as stated below):]
10.4.1	Language of arbitration	[Language]

3.9 施工管理	［包含在**服务**中的 / 未包含在**服务**中的］
7.4 对**咨询工程师（单位）**的第三方收费	［减免适用 / 减免不适用］
8.2 责任期限	
8.2.1 责任期限	［期限］
8.3 责任限度	
8.3.1 责任限度	［数额］
9 保险	
9.1.1 由**咨询工程师（单位）**承担的保险	
职业责任保险	［数额］
公共责任保险	［数额］
10 争端和仲裁⊖	

⊖ 此处中文按照勘误表改正后的英文翻译。——译者注

Part B Additional or Amended Clauses

The parties are to include in this section any variations, omissions and/or additions to the General Conditions.

B 部分　附加或修订的条款

双方应把对**通用条件**的变更、删减和 / 或增加编入本节。

APPENDICES

These Appendices form part of the Agreement.

1   **Scope of Services**

The following guidance is given to assist the parties to complete this appendix

Specify the scope of the Consultant's Services as finally negotiated and agreed – the description should be as comprehensive as reasonably practicable and, where beneficial to the understanding of the scope, should identify matters excluded from the scope. Clients/Consultants may refer to the FIDIC Definition of Services for guidance on preparing a scope of services.

Describe the function and purpose of the Services. Ensure that the function and purpose are consistent with the scope of Services and that such is described in terms that can be measured and verified. The Consultant must satisfy itself that the function and purpose is achievable using the standard of care in Sub-Clause 3.3.

Specify any information relied upon by the Consultant in the discharge of the Services that cannot be reviewed by the Consultant for accuracy and sufficiency under Sub-Clause 2.1.2, such as sub-surface or hydrological conditions.

Specify any Construction Administration requirements to be fulfilled by the Consultant including the form of Works Contract (e.g. FIDIC Red Book) under which the Consultant shall act.

Specify the responsibility for interface management between the Services and services provided by others (when the provision of services by others is necessary), if not the responsibility of the Client.

附录

这些**附录**构成本**协议书**的一部分。

1 **服务范围**

下列指南旨在帮助双方编写本附录

规定经最终协商和同意的**咨询工程师（单位）服务**的范围——描述应在合理可行的范围内尽可能全面，并在有助于理解范围的情况下，应明确范围之外的事项。**客户/咨询工程师（单位）**可参考菲迪克（FIDIC）**服务**定义中有关编写服务范围的指南。

描述**服务**的功能和目的。确保功能和目的与**服务**范围一致，并用可以衡量和验证的术语进行描述。**咨询工程师（单位）**必须保证，通过使用**第3.3款**中的谨慎标准是可以实现功能和目的的。

规定**咨询工程师（单位）**在提供**咨询工程师（单位）**不能评估**第2.1.2项**规定的**服务**的准确性和充分性时所依赖的任何信息，例如地下或水文条件。

规定**咨询工程师（单位）**要满足的任何**施工管理**要求，包括**咨询工程师（单位）**行事所依据的**工程合同**格式［例如，菲迪克（FIDIC）红皮书］。

如果不属于**客户**的责任，则规定**服务**与其他方提供的服务之间的接口管理责任（当需要由其他方提供服务时）。

2 Personnel, Equipment, Facilities and Services of Others to be Provided by the Client

The following guidance is given to assist the parties to complete this Appendix

List the requirements of personnel, equipment and facilities to be provided by the Client as completely and in as much detail as possible.

List and describe as completely, and in as much detail as possible, the services of others to be provided on behalf of the Client.

2 **要由客户提供的人员、设备、设施和其他方的服务**

*以下指南旨在帮助各方编写本**附录***

尽可能详尽地列出**客户**要提供的人员、设备和设施的要求。

尽可能详尽地列出和描述要代表**客户**提供的其他方的服务。

3 Remuneration and Payment

The following guidance is given to assist the parties to complete this Appendix

Appendix 3, as a minimum, should cover, as applicable:

- agreed remuneration, whether lump sum or schedule of rates or any combination thereof, to be paid to the Consultant for the performance of the Services
- terms of payment, percentage fees, timescale, lump sums
- rates and prices to be applied to Variations (where appropriate) or Exceptional Costs
- times for payment if not 28 days (Sub-Clause 7.2.1)
- process for submission of invoices and methods of payment
- price changes, inflation etc., if applicable.
- currencies of payment (Sub-Clause 7.3.1)
- financing charges – rate to be applied (Sub-Clause 7.2.2). The Parties should agree a rate that is meaningful within the commercial context of the Agreement
- taxation additional to payments (if any)
- allowable expenses

3　　　**报酬和付款**

下列指南旨在帮助各方编写本附录

如适用，**附录** 3 至少应包括：

- 应支付给**咨询工程师**（**单位**）履行**服务**的商定的报酬，无论是总价、费率表还是两者的组合；
- 支付条件、费用百分比、时间表、总价；
- 应用于**变更**（如适用）或**例外费用**的费率和价格；

- 付款次数，如非 28 天（**第 7.2.1 项**）；
- 提交发票的过程和付款方式；
- 价格变化、通货膨胀等，如适用；
- 支付货币（**第 7.3.1 项**）；
- 融资费用——要使用的费率（**第 7.2.2 项**），双方应商定在**协议书**的商业范围内有意义的费率；
- 付款以外的税费（如果有）；
- 允许的支出。

4   Programme

The following guidance is given to assist the parties to complete this Appendix

Appendix 4 should be used to expand upon the requirements for the Programme for the Services to be submitted under Clause 4.3.

The Appendix should identify the Commencement Date and completion date for the Services together with any other key dates for receipt or delivery of information between the Parties. Interface obligations with others should be noted here.

The Appendix should stipulate any requirements of the Client on the order or sequence of activities and any requirements of the Client for review and approval periods for the Services.

If required by the Client, any particular programme software to be used to produce the Programme should be stipulated here.

Information to be supplied by the Consultant to the Client on a monthly basis to report on progress against the Programme should be stipulated here.

4　　　　　进度计划

下列指南旨在帮助各方编写本附录

附录 4 应用于扩展要根据第 4.3 款提交**服务进度计划**的要求。

本**附录**应明确**服务**的**开始日期**和完成日期，以及双方之间接收或传递信息的任何其他关键日期。与其他方之间的接口义务应在此处注明。

本**附录**应规定**客户**对活动次序或顺序的任何要求，以及**客户**对**服务**的审核和批准期限的任何要求。

如果**客户**有要求，则应在此处规定用于生成**进度计划**的任何特定程序软件。

应规定**咨询工程师（单位）**每月向**客户**提供的信息，以报告该**进度计划**的进度。

5 Rules for Adjudication

General

1. Any reference in the Agreement to the Rules for Adjudication shall be deemed to be a reference to these Rules.

2. Definitions in the Agreement shall apply in these Rules.

Appointment of Adjudicator

3. The Parties shall jointly ensure the appointment of the Adjudicator. The Adjudicator shall be a suitably qualified person.

4. If for any reason the appointment of the Adjudicator is not agreed at the latest within 14 days of the reference of a dispute in accordance with these Rules, then either Party may apply, with a copy of the application to the other Party, to any appointing authority named in the Agreement or, if none, to the President of FIDIC or his nominee, to appoint an Adjudicator, and such appointment shall be final and conclusive.

5. The Adjudicator's appointment may be terminated by mutual agreement of the Parties. The Adjudicator's appointment shall expire when the Services have been completed or when any disputes referred to the Adjudicator shall have been withdrawn or decided, whichever is the later.

Terms of Appointment

6. The Adjudicator is to be, and is to remain throughout his appointment, impartial and independent of the Parties and shall immediately disclose in writing to the Parties anything of which he becomes aware which could affect his impartiality or independence.

7. The Adjudicator shall not give advice to the Parties or their representatives concerning the conduct of the project of which the Services form part other than in accordance with these Rules.

8. The Adjudicator shall not be called as a witness by the Parties to give evidence concerning any dispute in connection with, or arising out of, the Agreement.

9. The Adjudicator shall treat the details of the Agreement and all activities and hearings of the Adjudicator as confidential and shall not disclose the same without the prior written consent of the Parties. The Adjudicator shall not, without the consent of the Parties, assign or delegate any of his work under these Rules or engage legal or technical assistance.

10. The Adjudicator may resign by giving 28 days' notice to the Parties. In the event of resignation, death or incapacity, termination or a failure or refusal to perform the duties of Adjudicator under these Rules, the Parties shall agree upon a replacement Adjudicator within 14 days or Rule 4 shall apply.

11. The Adjudicator shall in no circumstances be liable for any claims for anything done or omitted in the discharge of the Adjudicator's duties unless the act or omission is shown to have been in bad faith.

12. If the Adjudicator shall knowingly breach any of the provisions of Rule 6 or act in bad faith, he shall not be entitled to any fees or expenses hereunder and shall reimburse each of the Parties for any fees and expenses properly paid to him if, as a consequence of such breach any proceedings or decisions of the Adjudicator are rendered void or ineffective.

5 裁决规则

概述

1 **协议书**中任何参照**裁决规则**的地方均应视为对**本规则**的参照。

2 **协议书**中的定义应适用于**本规则**。

裁决员的任命

3 双方应共同确保**裁决员**的任命。**裁决员**应为具有适当资格的人员。

4 如因任何原因,在根据**本规则**提出争端14天之内未就**裁决员**的任命达成商定,则任一方均可向另一方申请,并将申请书抄送另一方,任命**协议书**中指定的授权机构,或者,如果没有授权机构,则向菲迪克(FIDIC)主席或其指定人员,任命一名**裁决员**,该任命应是最终并具有决定性的。

5 经双方商定,**裁决员**的任命可以终止。**服务**已完成或任何提交给**裁决员**的争端应已撤回或裁定时,**裁决员**的任命应期满,以较晚者为准。

任命条件

6 **裁决员**应在整个任命期间保持公正和独立于**双方**,并应立即以书面形式向双方披露其了解到可能影响其公正或独立性的任何事项。

7 除遵守**本规则**外,**裁决员**不应就**服务**构成其一部分的项目的实施向双方或其代表提供建议。

8 双方不应要求**裁决员**作为证人,就与**协议书**有关或由**协议书**引起的任何争端提供证据。

9 **裁决员**应将**协议书**详情以及**裁决员**的所有活动和意见听取会内容视为机密,未经双方事先书面同意,不应披露。未经双方同意,**裁决员**不应将其在**本规则**规定的任何工作分配或委派给他人,也不得寻求法律或技术援助。

10 **裁决员**可提前28天向双方发出通知辞职。如果**裁决员**辞职、死亡或无行为能力、终止合同或未能或拒绝履行**本规则**规定的**裁决员**职责,则双方应在14天内商定更换**裁决员**,或**规则**4应适用。

11 除非证明该行为或遗漏是不诚信的,否则在任何情况下,**裁决员**均不应对在履行其职责过程中所做或遗漏的任何事项的任何索赔负责。

12 如果**裁决员**在知情的情况下违反了**规则**6的任何规定或不诚信行事,则其无权获得**协议书**规定的任何费用或开支,并且,如果因其违反规定,导致**裁决员**的任何诉讼或决定失效或无效,其应向双方退还已适当支付给他的任何费用和开支。

Payment

13　The Adjudicator shall be paid the fees and expenses set out in the Adjudicator's Agreement.

14　The retainer fee, if applicable, shall be payment in full for:

(a)　being available, on 28 days' notice, for all hearings and visits;
(b)　all office overhead expenses such as secretarial services, photocopying and office supplies incurred in connection with his duties;
(c)　all services performed hereunder except those performed during the days referred to in Rule 15.

15　The daily fee shall be payable for each working day preparing for or attending visits or hearings or preparing decisions including any associated travelling time.

16　The retainer and daily fees shall remain fixed for the period of tenure of the Adjudicator.

17　All payments to the Adjudicator shall be made by the Parties as determined by the Adjudicator. The Adjudicator's invoices for any monthly retainer shall be submitted quarterly in advance and invoices for daily fees and expenses shall be submitted following the conclusion of a visit or hearing. All invoices shall contain a brief description of the activities performed during the relevant period. The Adjudicator may suspend work if any invoice remains unpaid at the expiry of the period for payment, provided that 7 days prior notice has been given to both Parties.

18　If a Party fails to pay an invoice addressed to it, the other Party shall be entitled to pay the sum due to the Adjudicator and recover the sum paid from the defaulting Party.

Procedure for Obtaining Adjudicator's Decision

19　A dispute between the Parties may be referred in writing by either Party to the Adjudicator for his decision, with a copy to the other Party. If the Adjudicator has not been agreed or appointed, the dispute shall be referred in writing to the other Party, together with a proposal for the appointment of an Adjudicator. A reference shall identify the dispute and refer to these Rules.

20　The Adjudicator may decide to conduct a hearing in which event he shall decide on the date, place and duration for the hearing. The Adjudicator may request that written statements from the Parties be presented to him prior to, at or after the hearing. The Parties shall promptly provide the Adjudicator with sufficient copies of any documentation and information relevant to the Agreement that he may request.

21　The Adjudicator shall act as an impartial expert, not as an arbitrator, and shall have full authority to conduct any hearing as he thinks fit, not being bound by any rules or procedures other than those set out herein. Without limiting the foregoing, the Adjudicator shall have power to:

(a)　decide upon the Adjudicator's own jurisdiction, and as to the scope of any dispute referred to him,
(b)　make use of his own specialist knowledge, if any,
(c)　adopt an inquisitorial procedure,
(d)　decide upon the payment of interest in accordance with the Agreement,
(e)　open up, review and revise any opinion, instruction, determination, certificate or valuation, related to the dispute,

付款

13 应向**裁决员**支付**裁决员协议书**中规定的费用和开支。

14 如适用，聘用费应为全额支付的以下活动费用：

（a） 参加所有提前 28 天通知的意见听取会和考察的费用；
（b） 所有办公管理费用，如秘书服务、复印与其职责相关的办公用品的费用；
（c） **规则 15** 所述日期内履行的服务以外的所有服务的费用。

15 应按准备或参加考察或意见听取会，或编写决定的每个工作日支付日酬金，包括任何相关的旅行时间。

16 **裁决员**任职期间，聘用费和日酬金应保持不变。

17 给**裁决员**的所有付款均应由双方按**裁决员**确定的数额支付。**裁决员**月度聘用费的发票应提前每季度提交一次，日费和开支的发票应在考察或意见听取会结束后提交。所有发票均应包含在相关期间内进行的活动的简短描述。如果在付款期限届满前发票仍未得到支付，**裁决员**可在提前 7 天通知双方的前提下暂停工作。

18 如果一方未能支付提交给其的发票，则另一方应有权支付应付给**裁决员**的总额，并追回违约**方**支付的总额。

取得裁决员决定的程序

19 双方的争端可由任一方以书面形式提交给**裁决员**做出其决定，并抄送另一方。如果未商定或任命**裁决员**，则争端应以书面形式提交给另一方，并提出任命**裁决员**的建议。提交应明确说明争端并参照**本规则**。

20 **裁决员**可决定举行意见听取会，在此情况下，其应决定意见听取会的日期、地点和持续时间。**裁决员**可要求双方在意见听取会之前、期间或之后提交书面说明。双方应立即向**裁决员**提供其可能要求的、与**协议书**有关的任何文件和信息的足够副本。

21 **裁决员**应作为公正的专家而非仲裁员行事，并应有权在其认为合适的情况下进行任何意见听取会，不受**协议书**所列规则或程序以外的任何规则或程序的约束。在不限制上述情况的前提下，**裁决员**应有权：

（a） 决定**裁决员**自身的管辖权，以及委托给他的任何争端的范围；
（b） 利用其自身的专业知识，如果有；
（c） 采用研询程序；
（d） 根据**协议书**决定利息的支付；

（e） 公开、审查和修改与争端有关的任何意见、指示、确定、证明和估价；

(f) refuse admission to hearings to any persons other than the Client, the Consultant and their respective representatives, and to proceed in the absence of any Party who the Adjudicator is satisfied received notice of the hearing.

22 All communications between either of the Parties and the Adjudicator and all hearings shall be in the language of the Adjudicator's Agreement. All such communications shall be copied to the other Party.

23 No later than the fifty-sixth day after the day on which the Adjudicator received a reference or, if later, the day on which the Adjudicator's Agreement came into effect, the Adjudicator shall give written notice of his decision to the Parties. Such decision shall include reasons and state that it is given under these Rules.

(f) 拒绝**客户**、**咨询工程师**（**单位**）及其各自代表以外的任何人参加意见听取会，并在**裁决员**确认已收到意见听取会通知的任一方缺席的情况下继续进行会议。

22 双方中的任一方与**裁决员**之间的所有通信交流和所有意见听取会均应使用**裁决员协议书**的语言。所有此类通信交流均应抄送给另一方。

23 **裁决员**收到提交后不晚于第 56 天，或如果超过，则在**裁决员协议书**生效日，**裁决员**应将其决定书面通知双方。此类决定应包括理由并说明其是根据本**规则**做出的。

客户/咨询工程师（单位）服务协议书范本

Client/Consultant
MODEL SERVICES AGREEMENT

通用条件
General Conditions

协议书格式
FORM OF AGREEMENT

专用条件
PARTICULAR CONDITIONS

通用条件
GENERAL CONDITIONS

CONTENTS

General Conditions 30

1 GENERAL PROVISIONS 30

    1.1   Definitions
    1.2   Interpretation
    1.3   Notices and other Communications
    1.4   Law and Language
    1.5   Changes in Legislation
    1.6   Assignments and Sub-Contracts
    1.7   Intellectual Property
    1.8   Confidentiality
    1.9   Publication
    1.10 Anti-Corruption
    1.11 Relationship of Parties
    1.12 Agreement Amendment
    1.13 Severability
    1.14 Non Waiver
    1.15 Priority of Documents
    1.16 Good Faith

2 THE CLIENT 44

    2.1   Information
    2.2   Decisions
    2.3   Assistance
    2.4   Client's Financial Arrangements
    2.5   Supply of Client's Equipment and Facilities
    2.6   Supply of Client's Personnel
    2.7   Client's Representative
    2.8   Services of Others

3 THE CONSULTANT 48

    3.1   Scope of Services
    3.2   Function and Purpose of Services
    3.3   Standard of Care
    3.4   Client's Property
    3.5   Consultant's Personnel
    3.6   Consultant's Representative
    3.7   Changes in Consultant's Personnel
    3.8   Safety and Security of Consultant's Personnel
    3.9   Construction Administration

4 COMMENCEMENT AND COMPLETION 52

    4.1   Agreement Effective

目录

| | 通用条件 …………………………………………………………… | 31 |

1　一般规定 ………………………………………………………………… 31

    1.1　定义
    1.2　解释
    1.3　通知和其他通信交流
    1.4　法律和语言
    1.5　法律改变
    1.6　转让和分包
    1.7　知识产权
    1.8　保密
    1.9　出版
    1.10　反腐败
    1.11　双方关系
    1.12　协议书修改
    1.13　可分性
    1.14　不弃权
    1.15　文件的优先次序
    1.16　诚信

2　客户 …………………………………………………………………………… 45

    2.1　信息
    2.2　决定
    2.3　协助
    2.4　客户的资金安排
    2.5　客户设备和设施的提供
    2.6　客户人员的提供
    2.7　客户代表
    2.8　其他方的服务

3　咨询工程师（单位） ………………………………………………………… 49

    3.1　服务范围
    3.2　服务的功能和目的
    3.3　谨慎标准
    3.4　客户财产
    3.5　咨询工程师（单位）人员
    3.6　咨询工程师（单位）代表
    3.7　咨询工程师（单位）人员变动
    3.8　咨询工程师（单位）人员的安全和保障
    3.9　施工管理

4　开始和完成 …………………………………………………………………… 53

    4.1　协议书生效

        4.2   Commencement and Completion of Services
        4.3   Programme
        4.4   Delays
        4.5   Rate of Progress of Services
        4.6   Exceptional Event

5   VARIATIONS TO SERVICES 56

        5.1   Variations
        5.2   Agreement of Variation Value and Impact

6   SUSPENSION OF SERVICES AND TERMINATION OF AGREEMENT 58

        6.1   Suspension of Services
        6.2   Resumption of Suspended Services
        6.3   Effects of Suspension of the Services
        6.4   Termination of Agreement
        6.5   Effects of Termination
        6.6   Rights and Liabilities of the Parties

7   PAYMENT 64

        7.1   Payment to the Consultant
        7.2   Time for Payment
        7.3   Currencies of Payment
        7.4   Third-Party Charges on the Consultant
        7.5   Disputed Invoices
        7.6   Independent Audit

8   LIABILITIES 68

        8.1   Liability for Breach
        8.2   Duration of Liability
        8.3   Limit of Liability
        8.4   Exceptions

9   INSURANCE 70

        9.1   Insurances to be taken out by Consultant

10   DISPUTES AND ARBITRATION 72

        10.1   Amicable Dispute Resolution
        10.2   Adjudication
        10.3   Amicable Settlement
        10.4   Arbitration
        10.5   Failure to Comply with Adjudicator's Decision

4.2　服务的开始和完成
　　　4.3　进度计划
　　　4.4　延误
　　　4.5　服务进度
　　　4.6　例外事件

5　　服务变更 …………………………………………………………………… 57

　　　5.1　变更
　　　5.2　变更价值和影响协议书

6　　服务暂停和协议书终止 ……………………………………………………… 59

　　　6.1　服务暂停
　　　6.2　服务暂停的恢复
　　　6.3　服务暂停的影响
　　　6.4　协议书终止
　　　6.5　终止的影响
　　　6.6　双方权利和责任

7　　付款 …………………………………………………………………………… 65

　　　7.1　对咨询工程师（单位）的付款
　　　7.2　付款时间
　　　7.3　支付货币
　　　7.4　第三方对咨询工程师（单位）的收费
　　　7.5　有争议的发票
　　　7.6　独立审计

8　　责任 …………………………………………………………………………… 69

　　　8.1　违约责任
　　　8.2　责任期限
　　　8.3　责任限度
　　　8.4　例外情况

9　　保险 …………………………………………………………………………… 71

　　　9.1　由咨询工程师（单位）承担的保险

10　　争端和仲裁 …………………………………………………………………… 73

　　　10.1　友好解决争端
　　　10.2　裁决
　　　10.3　友好解决
　　　10.4　仲裁
　　　10.5　未能遵守裁决员的决定

General Conditions

1 General Provisions

1.1 Definitions

The following words and expressions shall have the meanings assigned to them except where the context otherwise requires:

1.1.1 "**Agreement**" means the Form of Agreement together with the Client/Consultant Model Services Agreement (General Conditions and Particular Conditions), Appendix 1 [*Scope of Services*], Appendix 2 [*Personnel, Equipment, Facilities and Services of Others to be Provided by the Client*], Appendix 3 [*Remuneration and Payment*], Appendix 4 [*Programme*] and Appendix 5 [*Rules for Adjudication*] and any letters of offer and acceptance attached to any of the above.

1.1.2 "**Background Intellectual Property**" means, in respect of each Party, the Intellectual Property owned by or otherwise in the possession of that Party at the Commencement Date.

1.1.3 "**Client**" means the Party named in the Form of Agreement and legal successors to the Client and permitted assignees.

1.1.4 "**Client's Representative**" means the person referred to in the Particular Conditions, or appointed from time to time by the Client, and communicated by Notice to the Consultant to be its representative for the administration of the Agreement.

1.1.5 "**Commencement Date**" means the date identified in the Particular Conditions; where no date is identified then the Commencement Date shall be 14 days after the Effective Date.

1.1.6 "**Confidential Information**" means all information specifically identified by the disclosing Party as confidential at the time of disclosure, or information that a reasonable person would consider from the nature of the said information and circumstances to be confidential, including without limitation confidential or proprietary information, trade secrets, data, documents, communications, plans, know-how, formulas, designs, calculations, test results, specimens, drawings, studies, specifications, surveys, photographs, software, processes, programmes, reports, maps, models, agreements, ideas, methods, discoveries, inventions, patents, concepts, research, development, and business and financial information.

1.1.7 "**Consultant**" means the professional firm or individual named in the Form of Agreement and legal successors to the Consultant and permitted assignees.

1.1.8 "**Consultant's Representative**" means the person referred to in the Particular Conditions or appointed from time to time by the Consultant, and communicated by Notice to the Client to be its representative for the administration of the Agreement.

1.1.9 "**Country**" means the country named in the Particular Conditions or, where no country is mentioned, the country where the Project site, or the main project site as the case may be, is located.

通用条件

1 一般规定

1.1
定义

除上下文另有要求外，下列词语和措辞应具有所赋予它们的含义：

1.1.1 "**协议书**"系指**协议书格式**以及**客户/咨询工程师（单位）**服务协议书范本（通用条件和专用条件）、附录1［*服务范围*］、附录2［*要由客户提供的人员、设备、设施和其他方的服务*］、附录3［*报酬和付款*］，附录4［*进度计划*］和附录5［*裁决规则*］以及附于上述任何一项的任何报价书和中标函。

1.1.2 "**背景知识产权**"系指就每一方而言，在**开始日期**该方拥有或以其他方式占有的**知识产权**。

1.1.3 "**客户**"系指**协议书格式**中命名的一方和**客户**的合法继承人以及许可受让人。

1.1.4 "**客户代表**"系指**专用条件**中提及的，或由**客户**不时任命的，并以**通知**形式通报**咨询工程师（单位）**作为其管理**协议书**的代表的人员。

1.1.5 "**开始日期**"系指**专用条件**中确定的日期；如果日期未确定，则**开始日期**应为**生效日期**后14天。

1.1.6 "**保密信息**"系指披露方在信息披露时明确确定为保密的所有信息，或一名合理人员根据所述信息的性质和情况认为是保密的信息，包括但不限于机密或专有信息、商业机密、数据、文件、通信交流、计划、专有技术、公式、设计、计算、测试结果、样本、图纸、探讨、规范、调查、照片、软件、流程、程序、报告、地图、模型、协议书、想法、方法、发现、发明、专利、概念、研究、开发、商业和资金信息。

1.1.7 "**咨询工程师（单位）**"系指**协议书格式**中命名的专业公司或个人，以及**咨询工程师（单位）**的合法继承人和许可受让人。

1.1.8 "**咨询工程师（单位）代表**"系指**专用条件**中提及的或**咨询工程师（单位）**不时任命的，并以**通知**形式通报**客户**作为其管理**协议书**的代表的人员。

1.1.9 "**工程所在国**"系指**专用条件**中指明的国家，或者，如果未提及任何国家，指项目现场或主要项目现场视情况而定，所在的国家。

1.1.10 "**day**" means a calendar day.

1.1.11 "**Effective Date**" means the date on which the Agreement comes into force and effect pursuant to Clause 4.1 [*Agreement Effective*].

1.1.12 "**Exceptional Costs**" means the costs, not otherwise compensated under the Agreement, arising out of any necessary work, cost, expense or delay incurred by the Consultant which is additional to the Services (or Variations) and which is necessarily and unavoidably performed under the Agreement and in each case identified as such in the Agreement.

1.1.13 "**Exceptional Event**" means an event or circumstance which is (a) beyond a Party's control; (b) which such Party could not reasonably have provided against before entering into the Agreement; (c) which having arisen, such Party could not reasonably have avoided or overcome; and (d) which is not substantially attributable to the other Party. An Exceptional Event may include, but is not limited to, events or circumstances of the kind listed below, subject to (a) to (d) above:

(i) war, hostilities (whether war be declared or not), invasion, act of foreign enemies;
(ii) rebellion, terrorism, revolution, insurrection, military or usurped power or civil war;
(iii) riot, commotion, disorder, strike or lockout by persons other than the Consultant's personnel and other employees of the Consultant and Consultant´s sub-consultants;
(iv) munitions of war, explosive materials, ionising radiation or contamination by radio-activity except as may be attributable to the Consultant's actions;
(v) natural catastrophes such as earthquake, hurricane, typhoon or volcanic activity.

1.1.14 "**Foreground Intellectual Property**" means all Intellectual Property created as a result of the Services performed by the Consultant.

1.1.15 "**Form of Agreement**" means the document entitled Form of Agreement which forms part of the Agreement.

1.1.16 "**FIDIC**" means the Fédération Internationale des Ingénieurs-Conseils, the International Federation of Consulting Engineers.

1.1.17 "**Intellectual Property**" means all intellectual property rights including, without limitation, any patents, patent application, trademarks, trade secrets, registered designs, registered design application, copyrights, design rights, moral rights, process, formula, specification, drawing, including rights in computer software and databases howsoever arising in any part of the world.

1.1.18 "**Local Currency**" means the currency of the Country and "**Foreign Currency**" means any other currency.

1.1.19 "**Notice**" means a written communication identified as a Notice and issued in accordance with the provisions of Clause 1.3 [*Notices and other Communications*].

1.1.20 "**Party**" and "**Parties**" means the Client and/or the Consultant as the context requires.

1.1.10 "天"系指一个日历日。

1.1.11 "生效日期"系指**协议书**根据第4.1款[**协议书生效**]的规定生效并实施的日期。

1.1.12 "例外费用"系指**咨询工程师（单位）**产生的因任何必要的工作、成本，以及支出或延误而发生的，且**协议书**中未予补偿的费用，是**服务**（或**变更**）的附加费用，是**协议书**规定有必要和不可避免地履行的，并且在每种情况下都在**协议书**中明确的。

1.1.13 "例外事件"系指（a）超出一方控制范围的；（b）该方在签订**协议书**之前无法合理预防的；（c）发生后，该方无法合理避免或克服的；以及（d）实质上不可归因于另一方的事件或情况。根据上述（a）至（d）段，**例外事件**可能包括但不限于下列事件或情况：

（i） 战争、敌对行动（不论宣战与否）、入侵、外敌行为；

（ii） 叛乱、恐怖主义、革命、暴动、军事或篡权或内战；

（iii） **咨询工程师（单位）**人员和**咨询工程师（单位）**的其他雇员以及**咨询工程师（单位）**的分包咨询工程师以外的人员的暴乱、骚乱、混乱、罢工或停工；

（iv） 军需品、爆炸性材料、电离辐射或放射性污染，但由**咨询工程师（单位）**的行为引起的情况除外；

（v） 自然灾害，如地震、飓风、台风或火山活动。

1.1.14 "前景知识产权"系指**咨询工程师（单位）**履行的**服务**所产生的所有知识产权。

1.1.15 "协议书格式"系指作为构成**协议书**一部分的名为**协议书格式**的文件。

1.1.16 "菲迪克（FIDIC）"系指国际咨询工程师联合会。

1.1.17 "知识产权"系指所有知识产权，包括但不限于任何专利、专利申请、商标、商业秘密、注册外观设计、注册外观设计申请、版权、设计权、精神权利、工艺、配方、规范、图纸，包括在世界任何地方产生的计算机软件和数据库的权利。

1.1.18 "当地货币"系指工程所在国的货币，"外国货币"系指任何其他货币。

1.1.19 "通知"系指根据第1.3款[**通知和其他通信交流**]的规定签发的书面**通知**。

1.1.20 "一方"和"双方"系指**客户**和/或**咨询工程师（单位）**，视上下文而定。

1.1.21 "**Programme**" shall have the meaning given to it in Clause 4.3 [*Programme*].

1.1.22 "**Project**" means the project named in the Particular Conditions for which the Services are to be provided.

1.1.23 "**Services**" means the services defined in Appendix 1 [*Scope of Services*] to be performed by the Consultant in accordance with the Agreement which includes any Variations to the Services instructed or arising in accordance with the Agreement.

1.1.24 "**Time for Completion**" means the time for completing the Services as stated in the Particular Conditions, or as may be amended in accordance with the Agreement, calculated from the Commencement Date.

1.1.25 "**Variation**" or "**Variation to the Services**" means any change to the Services instructed or approved as a Variation under Clause 5.1 [*Variations*].

1.1.26 "**Variation Notice**" means a written communication identified as a Variation Notice and issued in accordance with the provisions of Clause 1.3 [*Notices and other Communications*]

1.1.27 "**Works Contract**" means a contract for the performance of permanent and temporary works (if any) to be carried out by a contractor appointed by the Client for the achievement of the Project.

1.1.28 "**year**" means a calendar year.

1.2 Interpretation

1.2.1 Words indicating the singular include the plural, and vice-versa where the context requires.

1.2.2 Words indicating one gender include all genders.

1.2.3 Provisions including the word "**agree**", "**agreed**" or "**agreement**" require the agreement to be recorded in writing, and signed by both Parties.

1.2.4 "**shall**" means that the Party or person referred to has the obligation under the Agreement to perform the duty referred to.

1.2.5 "**may**" means that the Party or person referred to has the choice of whether to act or not in the matter referred to.

1.2.6 "**written**" or "**in writing**" means hand written, type-written, printed or electronically made and resulting in a permanent uneditable record.

1.2.7 any reference to "**price**", "**rates**", "**costs**", "**expenses**", "**damages**" and the like shall be a reference to the value of such item net of any applicable taxes unless specified otherwise.

1.3 Notices and other Communications

1.3.1 Wherever the Agreement provides for the giving or issuing of a Notice, a Variation Notice or other form of communication including without limitation approvals, consents, instructions and decisions, then such Notice, Variation Notice or communication shall be:

(a) where it is a Notice or Variation Notice, identified as such with reference to the Clause or Sub-Clause under which it is issued;

	1.1.21	"进度计划"的含义见第4.3款[*进度计划*]。
	1.1.22	"项目"系指在**专用条件**中指明的将提供**服务**的项目。
	1.1.23	"服务"系指附录1[*服务范围*]中规定的由**咨询工程师（单位）**根据**协议书**履行的服务，根据指示的或根据**协议书**产生的任何**服务变更**。
	1.1.24	"**完成时间**"系指**专用条件**中规定的，或根据**协议书**可能修改的、从**开始日期**算起的完成**服务**的时间。
	1.1.25	"变更"或"服务变更"系指根据第5.1款[*变更*]的规定对**服务**指示或批准为**变更**的任何改变。
	1.1.26	"变更通知"系指根据第1.3款[*通知和其他通信交流*]的规定，确定为**变更通知**并签发的书面通信交流。
	1.1.27	"工程合同"系指由**客户**为实现**项目**而指定的由承包商实施的执行永久和临时性工程（如果有）的合同。
	1.1.28	"年"系指日历年。

1.2 解释	1.2.1	表示单数的词语也包括复数，如果上下文需要，反之亦然。
	1.2.2	表示一种性别的词语包括所有性别。
	1.2.3	包括"同意""商定的"或"协议书"等字样的条款要求以书面形式记录协议，并由双方签字。
	1.2.4	"应当"系指提及的一方或个人有义务根据**协议书**履行所述义务。
	1.2.5	"可以"系指提及的一方或个人可以就所述事项选择作为或不作为。
	1.2.6	"书面"或"以书面"系指手写、打字、印刷或电子制作，并形成永久不可编辑的记录。
	1.2.7	除非另有规定，"价格""费率""成本""费用""损害赔偿费"等应指该事项扣除任何适用税款后的价值。

1.3 通知和其他通信交流	1.3.1	如果**协议书**对**通知**的发出或签发、**变更通知**或其他形式的通信交流做了规定，包括但不限于批准、同意、指示和决定，则此类**通知**、**变更通知**或通信交流应：
		（a） 如果是**通知**或**变更通知**，参考其签发所依据的**条**或**款**予以确定；

(b) where it is another form of communication, identified as such with reference to the Clause or Sub-Clause under which it is issued where appropriate;

(c) in writing and delivered by hand (against receipt), sent by mail or courier, or transmitted by any form of agreed system of electronic transmission stated in the Particular Conditions; and

(d) delivered, sent or transmitted to the address for the recipient's communications as stated in the Particular Conditions. However:

(i) if the recipient gives Notice of another address, Notices and other forms of communication shall thereafter be delivered accordingly; and

(ii) if the recipient has not stated otherwise when requesting an approval or consent, it may be sent to the address from which the request was issued.

Notices and other form of communications shall not be unreasonably withheld or delayed.

1.4 Law and Language

1.4.1 The Agreement shall be governed by the law stated in the Particular Conditions or, if no governing law is stated in the Particular Conditions, by the law of the Country.

1.4.2 If any part of the Agreement is written in more than one language then the ruling language shall be that stated in the Particular Conditions.

1.4.3 The language for all communications shall be the ruling language stated in the Particular Conditions or where no language is stated then all communications shall be in the language in which the Agreement (or most of it) is written.

1.5 Changes in Legislation

1.5.1 If after the date of the Consultant's offer/proposal in relation to the Agreement the scope, extent, nature or type of Services is affected by any change to national (or state) legislation, any statute, statutory instrument, order, regulation, bylaw, code or other legislation having application to the Services then such change to the Services shall be treated as a Variation to the Services under Clause 5.1 [*Variations*].

1.5.2 If after the date of the Consultant's offer/proposal in relation to the Agreement any change to national (or state) legislation, any statute, statutory instrument, order, regulation, bylaw, code or other legislation in any country in which the services are required by the Client, causes the Consultant to incur Exceptional Costs, then the agreed remuneration shall be adjusted in accordance with Sub-Clause 7.1.2 [*Payment to the Consultant*], and the Time for Completion amended in accordance with Clause 4.4 [*Delays*]. As soon as reasonably practicable the Consultant shall inform the Client by issue of a Notice of the occurrence of the Exceptional Costs.

1.6 Assignments and Sub-Contracts

1.6.1 Neither the Client nor the Consultant shall at any time assign the benefit of the Agreement without the prior written consent of the other. Such consent shall not be unreasonably withheld or delayed.

1.6.2 Neither the Client nor the Consultant shall assign obligations under the Agreement without the written consent of the other Party.

（b） 如果是另一种形式的通信交流，在适当情况下，参考其签发所依据的**条目**或**条款**予以确定；

（c） 以书面形式并由人面交（取得对方收据）、通过邮寄或信差传送，或通过**专用条件**中规定的任何商定的电子传输系统发送；（以及）

（d） 交付、传送或传输至**专用条件**中规定的收件人通信地址。但是：

（i） 如接收人**通知**了另外地址，**通知**和其他形式的通信交流应随后相应交付；（以及）

（ii） 如接收人在请求批准或同意时未另外说明，可以将其发送至签发请求的地址。

通知和其他形式的通信交流不得无故被扣压或延迟。

1.4 法律和语言

1.4.1 **协议书**应受**专用条件**规定的法律管辖，或者，如果**专用条件**未规定适用法律，则受**工程所在国**的法律管辖。

1.4.2 如果**协议书**的任何部分的文本采用一种以上的语言编写，则主导语言应为**专用条件**中规定的语言。

1.4.3 所有通信交流的语言应为**专用条件**中规定的主导语言，如未规定语言，则所有通信交流应使用**协议书**（或其大部分）的编写语言。

1.5 法律改变

1.5.1 如果在**咨询工程师（单位）**提出与**协议书**相关的报价/建议书之日后，**服务**范围、规模、性质或类型受到国家（或州）立法、任何法规、法定文书、命令、条例、细则、规范或适用于**服务**的其他立法的任何改变的影响，则此类**服务**的改变应视为第 5.1 款［**变更**］规定的**服务变更**。

1.5.2 如果在**咨询工程师（单位）**提出与**协议书**相关的报价/建议书之日后，任何国家（或州）立法、任何法规、法定文书、命令、条例、细则、规范或**客户**要求的**服务**⊖的任何国家的其他立法的任何改变，导致**咨询工程师（单位）**招致**例外费用**，则商定的报酬应根据第 7.1.2 项［对**咨询工程师（单位）**付款］的规定进行调整，并根据第 4.4 款［**延误**］的规定修改完成时间。**咨询工程师（单位）**应在合理可行的情况下尽快通过签发**通知**告知**客户**产生**例外费用**的情况。

1.5.3 一方可通过给另一方单独发出**通知**，要求修改**协议书**的规定，以符合适用法律改变⊖。

1.6 转让和分包

1.6.1 未经另一方事先书面同意，**客户**和**咨询工程师（单位）**任何时候均不得转让**协议书**的利益。不得无故拒绝或延迟此类同意。

1.6.2 未经另一方书面同意，**客户**和**咨询工程师（单位）**均不得转让**协议书**规定的义务。

⊖ 此处中文按照勘误表改正后的英文翻译。——译者注

1.6.3 The Consultant shall not sub-contract performance of all or part of the Services without the written consent of the Client. The consent of the Client shall not be required where the appointment of a sub-consultant for the performance of part of the Services is included in the Consultant's offer/proposal, if any, as incorporated into the Agreement, or is otherwise anticipated in any of the documents constituting the Agreement.

1.6.4 The Client's consent to any sub-contract arrangement shall not relieve the Consultant of any of the Consultant's obligations under the Agreement. The Consultant shall remain responsible and liable to the Client for the acts, omissions and defaults of the sub-consultant in relation to the Agreement as if they were acts, omissions and defaults of the Consultant.

1.7 Intellectual Property

1.7.1 All Intellectual Property held in any medium, whether electronic or otherwise, created by the Consultant during the performance of the Services (Foreground Intellectual Property) shall be vested in the Consultant. The Consultant shall grant to the Client a royalty free worldwide licence to use and copy the Foreground Intellectual Property for any purpose in connection with the Project.

1.7.2 All Background Intellectual Property shall remain the property of the original owner. The Consultant hereby grants to the Client, or agrees to procure the grant to the Client of an unrestricted royalty free licence to use and copy the Consultant's Background Intellectual Property to the extent reasonably required to enable the Client to make use of the Services or the Project. The Client hereby grants to the Consultant an unrestricted royalty free licence to use and copy the Client's Background Intellectual Property provided to the Consultant to the extent reasonably required to enable the Consultant to provide the Services.

1.7.3 The Consultant shall ensure (except in respect of any of the Client's Background Intellectual Property) that the Foreground Intellectual Property and the Consultant's Background Intellectual Property, to the extent incorporated into the Services, will not infringe any Intellectual Property or other rights of any third party.

1.7.4 The Consultant shall not be liable for the use by any person of the Consultant's Background Intellectual Property or the Consultant's Foreground Intellectual Property for any purpose other than the purpose for which it was originally intended.

1.7.5 In the event that the Client is in default of payment of any amounts due under the Agreement then the Consultant may upon seven (7) days' Notice revoke any licence granted therein.

1.8 Confidentiality

1.8.1 Except with the prior written consent of the other Party, neither Party shall disclose or cause or permit their employees, professional advisers, agents or sub-consultants to disclose to third parties any Confidential Information.

1.8.2 The restrictions on use and disclosure set forth in Sub-Clause 1.8.1 shall not apply to any information:

(a) which at the date of its disclosure is public knowledge or which subsequently becomes public knowledge other than by any act or failure to act on the part of the receiving Party or persons for whom the receiving Party has assumed responsibility under the Agreement;

	1.6.3	未经**客户**书面同意，**咨询工程师（单位）**不应将全部或部分**服务**分包出去。如果**咨询工程师（单位）**的报价/建议书中包含了为履行部分**服务**而任命的分包咨询工程师，如果有，纳入了**协议书**中，或在构成**协议书**的任何文件中另有预期，则无须征得**客户**的同意。
	1.6.4	**客户**对任何分包合同安排的同意不应免除**协议书**规定的**咨询工程师（单位）**的任何义务。**咨询工程师（单位）**应对分包咨询工程师与**协议书**有关的行为、遗漏和违约向**客户**负责，如同这些是**咨询工程师（单位）**的行为、遗漏和违约。
1.7 知识产权		
	1.7.1	**咨询工程师（单位）**在履行**服务**（前景知识产权）期间创造的任何成果，无论是电子的还是其他方式的，其所有**知识产权**应归属于**咨询工程师（单位）**。**咨询工程师（单位）**应授予**客户**免版税证书，供其为实现**项目**在全球范围内使用和复制**前景知识产权**。
	1.7.2	所有**背景知识产权**仍为原所有人的财产。**咨询工程师（单位）**特此向**客户**授予或同意向**客户**授予无限免版税证书，在合理需要的情况下使用和复制**咨询工程师（单位）**的**背景知识产权**，使**客户**能够使用于**服务**或**项目**。**客户**特此授予**咨询工程师（单位）**无限免版税证书，在合理需要的情况下使用和复制提供给**咨询工程师（单位）**的**客户**的**背景知识产权**，以使**咨询工程师（单位）**能够提供于**服务**。
	1.7.3	**咨询工程师（单位）**应确保（与任何**客户**的**背景知识产权**有关的情况除外）纳入**服务**范围的**前景知识产权**和**咨询工程师（单位）**的**背景知识产权**，不会侵犯任何第三方的任何**知识产权**或其他权利。
	1.7.4	对任何人将**咨询工程师（单位）**的**背景知识产权**或**咨询工程师（单位）**的**前景知识产权**用于其最初设定目的以外的任何其他目的，**咨询工程师（单位）**概不负责。
	1.7.5	如果**客户**未能支付**协议书**规定的任何应付款额，**咨询工程师（单位）**可提前七（7）天发出**通知**，撤销在此授予的任何证书。
1.8 保密		
	1.8.1	未经另一方事先书面同意，任一方均不得向第三方披露或促使或允许其雇员、专业顾问、代理人或分包咨询工程师向第三方披露任何保密信息。
	1.8.2	第 1.8.1 项规定的使用和披露限制对以下任何信息均应不适用：
		（a） 在披露之日已为公众所知或随后成为公众所知的信息，但代表接收方或接收方对其承担**协议书**规定的责任的人员的任何作为或不作为除外；

(b) which the receiving Party can establish by written proof was already in its possession at the time of disclosure by the disclosing Party and was not acquired directly or indirectly from the disclosing Party;

(c) which at any time after the Commencement Date has been acquired from any third party who did not acquire such information directly or indirectly from the disclosing Party or any of the disclosing Party's employees or professional advisers;

(d) which by proof in writing has been independently developed by the receiving Party without the use of Confidential Information; or

(e) which is required to be disclosed by law or order of a court of competent jurisdiction or government, department, agency or other public authority.

1.8.3 The obligations set forth in Sub-Clause 1.8.1 shall expire two (2) years after completion of the Services or the termination of the Agreement (whichever is the earlier) unless stated otherwise in the Particular Conditions.

1.9 Publication

1.9.1 Subject to Clause 1.8 [*Confidentiality*] and unless otherwise specified in the Particular Conditions, the Consultant, either alone or jointly with others, may publish material relating to the Services. Publication shall be subject to approval of the Client if it is within two (2) years of completion of the Services or termination of the Agreement (whichever is the earlier).

1.9.2 The Consultant may use material and information relating to the Services and the Project for commercial tendering purposes.

1.10 Anti-Corruption

1.10.1 In the performance of their obligations under the Agreement, the Consultant and the Client, their agents and employees shall comply with all applicable laws, rules, regulations and orders of any applicable jurisdiction, including without limitation those relating to corruption and bribery. The Parties shall also comply with the standards provided in the OECD Convention on Combating Bribery of Foreign Public Officials in International Business Transactions.

The Consultant hereby represents, warrants and covenants that:

a) it shall not participate, directly or indirectly in bribery, extortion, fraud, deception, collusion, cartels, abuse of power, embezzlement, trading in influence, money laundering, use of insider information, the possession of illegally obtained information or any other criminal activity; and

b) it shall neither receive nor offer, pay or promise to pay either directly or indirectly, anything of value to a "public official" (as defined below) in connection with any business opportunities which are the subject of the Agreement. Furthermore, the Consultant shall immediately give Notice to the Client with full particulars in the event that the Consultant receives a request from any public official requesting illicit payments.

1.10.2 A "public official" is:

(a) any official or employee of any government agency or government-owned or controlled enterprise;

（b）接收方可通过书面证据证明，在披露方披露时，该信息已为其所有，且不是直接或间接从披露方获得的信息；

（c）在**开始日期**后的任何时间，从任何第三方获得的信息，该第三方未直接或间接从披露方或披露方的任何雇员或专业顾问处获得此类信息；

（d）通过书面证据，接收方在未使用**保密信息**的情况下独立开发的信息；（或）

（e）有管辖权的法院或政府、部门、机构或其他公共机构的法律或命令要求披露的信息。

1.8.3 　除非**专用条件**中另有规定，**第 1.8.1 项**规定的义务应在**服务**完成或**协议书**终止后两（2）年内到期（以较早者为准）。

1.9 出版

1.9.1 　根据**第 1.8 款**[**保密**]的规定，除非**专用条件**中另有规定，**咨询工程师（单位）**可单独或与其他人联合出版与**服务**有关的资料。如果是在**服务**完成或**协议书**终止后的两（2）年内出版（以较早者为准），则应征得**客户**的批准。

1.9.2 　**咨询工程师（单位）**可将与**服务**和**项目**有关的资料和信息用于商务投标。

1.10 反腐败

1.10.1 　在履行**协议书**规定的义务时，**咨询工程师（单位）**和**客户**、其代理人和雇员均应遵守任何适用司法管辖区的所有适用法律、法规、条例和命令，包括但不限于与腐败和贿赂有关的法律、法规、条例和命令。双方还应遵守**经合组织《打击国际商业交易中行贿外国公职人员行为的公约》**规定的标准。

咨询工程师（单位）特此声明、保证和承诺：

a) 不直接或间接参与贿赂、勒索、欺诈、欺骗、串通、联盟、滥用职权、贪污、影响交易、洗钱、利用内幕信息、持有非法获取的信息或任何其他犯罪活动；（以及）

b) 不直接或间接向"公职人员"（定义见下文）收取、提供、支付或承诺支付与构成**协议书**主题任何商业机会有关的任何有价物品。此外，如果**咨询工程师（单位）**收到任何公职人员要求非法付款的请求，**咨询工程师（单位）**应立即**通知客户**并提供充分详细证明资料。

1.10.2 　"公职人员"系指：

（a）任何政府机构或国有或控股企业的任何官员或雇员；

(b) any person performing a public function;
(c) any official or employee of a public international organization including without limitation donor or funding agencies or the Client;
(d) any candidate for political office; or
(e) any political party or an official of a political party.

1.10.3 In conjunction with the requirements of this Clause 1.10 the Consultant shall at the Client's request demonstrate that it adheres to a documented code of conduct in respect to the prevention of corruption and bribery. As a minimum the Consultant shall comply with the FIDIC Code of Ethics and the FIDIC Integrity Management System available at www.fidic.org.

1.11 Relationship of Parties

1.11.1 Nothing contained in the Agreement shall be construed as creating a partnership, agency or joint venture between the Parties.

1.11.2 Where either Party consists of a joint venture or consortium then members of such joint venture or consortium shall be jointly and severally liable under the Agreement.

1.12 Agreement Amendment

1.12.1 The Agreement can only be amended with the written agreement of the Parties.

1.13 Severability

1.13.1 If any term or provision under the Agreement is held to be illegal or unenforceable in whole or in part then such term or provision shall be disregarded without affecting the enforceability of the remainder of the Agreement. Where either Party cannot rely on any term or provision, the Parties will negotiate in good faith for an alternative term or provision with similar contractual effect for both Parties.

1.14 Non Waiver

1.14.1 No failure or delay by either Party in exercising any of its rights under the Agreement shall operate as a waiver of such rights. Any waiver given by either Party in connection with the Agreement is binding only if it is served as a Notice and then strictly in accordance with the terms of the Notice.

1.15 Priority of Documents

1.15.1 The documents forming the Agreement are to be taken as mutually explanatory of one another. If there is a conflict between these documents then the documents shall be interpreted and construed in accordance with the order of precedence of documents given in Clause 2 of the Form of Agreement. If the conflict cannot be so resolved then the Client shall issue an instruction or Variation to the Services under Clause 5.1 [Variations] as the case may require, in order to resolve the conflict.

1.16 Good Faith

1.16.1 In all dealings under the Agreement the Client and the Consultant shall act in good faith and in a spirit of mutual trust.

　　　　　　　　　　（b）　任何履行公共职能的人；
　　　　　　　　　　（c）　国际公共组织的任何官员或雇员，包括但不限于捐赠者或资助机构或**客户**；

　　　　　　　　　　（d）　任何政治职位候选人；（或）
　　　　　　　　　　（e）　任何政党或政党的官员。

　　　　　　　　1.10.3　根据**第 1.10 款**的要求，**咨询工程师（单位）**应根据**客户**要求，证明其遵守了与防止腐败和贿赂有关的行为规范。作为最低要求，**咨询工程师（单位）**应遵守**菲迪克（FIDIC）道德规范和菲迪克（FIDIC）廉洁管理体系**，请浏览 www.fidic.org。

1.11 双方关系

　　　　　　　　1.11.1　**协议书**包含的任何内容不得解释为在双方之间建立合伙企业、代理机构或联营体。

　　　　　　　　1.11.2　如果任一方由联营体或财团组成，则该联营体或财团的成员应承担**协议书**规定的共同的和各自的责任。

1.12 协议书修改

　　　　　　　　1.12.1　**协议书**只有在双方书面商定的情况下才能修改。

1.13 可分性

　　　　　　　　1.13.1　如果**协议书**规定的任何条款或规定被认定为全部或部分不合法或不可执行，则在不影响**协议书**其余部分的可执行性的情况下，应忽略该条款或规定。如果任一方不能依赖任何条款或规定，双方应本着诚信原则协商一个对双方具有类似合同效力的替代条款或规定。

1.14 不弃权

　　　　　　　　1.14.1　任一方未能行事或拖延行使其**协议书**规定的任何权利，不得视为对该权利的放弃。任一方与**协议书**相关的任何弃权，只有在作为**通知**送达并严格遵守**通知**条款的情况下才具有约束力。

1.15 文件的优先次序

　　　　　　　　1.15.1　构成**协议书**的文件应视为相互解释说明的。如果此类文件之间存在冲突，则应按照**协议书格式第 2 条**中给出的文件优先次序对文件进行解释和说明。如果无法解决冲突，则**客户**应视情况需要，根据**第 5.1 款 [变更]** 对**服务**签发指示或**变更**，以解决冲突。

1.16 诚信

　　　　　　　　1.16.1　在**协议书**规定的所有交易中，**客户和咨询工程师（单位）**应本着真诚和相互信任的精神行事。

The Client

2.1 Information

2.1.1 In order not to delay the Consultant in the performance of the Services, the Client shall within a reasonable time and with due regard to the Programme, provide to the Consultant, free of cost, all information, and any further information reasonably requested by the Consultant, which may pertain to the Services and which the Client is able to obtain.

2.1.2 The Client accepts responsibility for and acknowledges that the Consultant will rely on the accuracy, sufficiency and consistency of all the information provided by the Client or by others on behalf of the Client. The Consultant shall use reasonable endeavours to review all significant information provided to it by the Client or by others on behalf of the Client within a reasonable time of receipt. To the extent achievable using the Standard of Care in Sub-clause 3.3.1 [*Standard of Care*], the Consultant shall review such information with a view to ensuring that such information does not contain any manifest error, omission or ambiguity and shall give Notice to the Client promptly of any adverse findings.

2.1.3 In the event of any error, omission, or ambiguity (for the avoidance of doubt, including a manifest error, omission or ambiguity) in the information provided to the Consultant then the Client shall rectify such matter by Notice and where necessary shall issue a Variation to the Services in accordance with Clause 5.1 [*Variations*] as the case may require.

2.2 Decisions

2.2.1 On all matters properly referred to the Client in writing by the Consultant, the Client shall give its decision, approval, consent, instruction or Variation, as the case may be, in writing within a reasonable time and with regard to the Programme so as not to delay the Services.

2.3 Assistance

2.3.1 In the Country and in respect of the Consultant, its personnel and dependants, as well as sub-consultants, if any, as the case may be, the Client shall do all in its power to assist in:

(a) the provision of documents necessary for entry, residency, working and exit;
(b) providing unobstructed access wherever it is required for the Services;
(c) import, export and customs clearance of personal effects and of goods required for the Services;
(d) their repatriation in emergencies;
(e) the provision of the authority necessary for the Consultant to permit the import of foreign currency by the Consultant for the Services and by its personnel for their personal use and to permit the export of money earned in the performance of the Services; and
(f) providing access to other organisations for collection of information which is to be obtained by the Consultant.

Sub-Clauses 2.3.1 (a) and (c) to (e) shall not apply where the Country is a principal place of business of the Consultant.

2 客户

2.1 信息

2.1.1 为了不延误**咨询工程师（单位）**提供**服务**的时间，**客户**应在合理的时间内，并在适当考虑**进度计划**的情况下，免费向**咨询工程师（单位）**提供**咨询工程师（单位）**合理要求的可能与**服务**有关的、**客户**能够获得的所有信息和任何进一步信息。

2.1.2 **客户**对**客户**或代表**客户**的其他人员提供的所有信息的准确性、充分性和一致性负责，并保证**咨询工程师（单位）**可信赖这些信息的准确性、充分性和一致性。**咨询工程师（单位）**应尽合理努力在收到**客户**或代表**客户**的其他人员向其提供的所有重要信息后的合理时间内进行审核。在使用第 3.3.1 项 [**谨慎标准**] 中规定的谨慎标准[注]可实现的范围内，**咨询工程师（单位）**应审核此类信息，以确保此类信息不包含任何明显的错误、遗漏或歧义，并应立即将发现的任何错误**通知客户**。

2.1.3 如果向**咨询工程师（单位）**提供的信息中存在任何错误、遗漏或歧义（为避免疑问，包括明显的错误、遗漏或歧义），则**客户**应通过**通知**纠正此类事项，必要时应根据**第 5.1 款**[**变更**]的规定视情况需要签发**服务变更**。

2.2 决定

2.2.1 对于**咨询工程师（单位）**以书面形式适当提交给**客户**的所有事项，**客户**应在合理时间内视情况以书面形式就**进度计划**做出决定、批准、同意、指示或**变更**，以免延误**服务**。

2.3 协助

2.3.1 在**工程所在国**，对于**咨询工程师（单位）**、其人员和家属以及分包咨询工程师，如果有，视情况而定，**客户**应尽其所能协助：

（a） 提供入境、居留、工作和出境所需的文件；

（b） 在**服务**所需的任何地方提供畅通无阻的通道；

（c） 个人物品和**服务**所需货物的进出口和清关；

（d） 紧急情况下的遣返；

（e） 为**咨询工程师（单位）**提供必要的授权，以允许**咨询工程师（单位）**为**服务**及其人员进口个人使用的外币，并允许出口在履行**服务**过程中获得的资金；（以及）

（f） 提供**咨询工程师（单位）**获取其他组织信息所需的渠道。

如果**工程所在国**是**咨询工程师（单位）**的主要营业地，则第 2.3.1 项的（a）段和（c）至（e）段应不适用。

⊖ 此处中文按照勘误表改正后的英文翻译。——译者注

2.4 Client's Financial Arrangements

2.4.1 The Client shall submit to the Consultant, within twenty eight (28) days of receiving any request from the Consultant, reasonable evidence that financial arrangements have been made and are being maintained which will enable the Client to make timely payments under Appendix 3 [*Remuneration and Payment*] or any other provision of the Agreement.

2.4.2 If the Client intends to make any material change to its financial arrangements, the Client shall give Notice to the Consultant with supporting particulars. In the event that the Consultant, acting reasonably, is not satisfied with the proposed change and/or detailed particulars submitted by the Client, then the Consultant shall be entitled to suspend the Services pursuant to Sub-Clause 6.1.2 (c) [*Suspension of Services*].

2.5 Supply of Client's Equipment and Facilities

2.5.1 The Client shall make the equipment and facilities described in Appendix 2 [*Personnel, Equipment, Facilities and Services of Others to be Provided by the Client*] available to the Consultant for the purpose of the Services, with due regard to the Programme and free of cost.

2.6 Supply of Client's Personnel

2.6.1 In consultation with the Consultant, the Client shall at its own cost arrange for the selection and provision of suitably qualified personnel in its employment to the Consultant in accordance with the requirements, if any, in Appendix 2 [*Personnel, Equipment, Facilities and Services of Others to be Provided by the Client*]. In connection with the provision of the Services and subject to the requirements of the applicable law(s), such personnel shall take instructions from the Consultant only.

2.6.2 The personnel to be supplied by the Client, and any future replacements that may be necessary, shall be subject to acceptance by the Consultant. Such acceptance shall not be unreasonably withheld.

2.6.3 In the event that the Client cannot supply the Client's personnel for which it is responsible or the performance of the Client's personnel supplied to the Consultant is not, in the reasonable opinion of the Consultant, adequate to discharge the services assigned to them, then the Consultant shall arrange for an alternative supply of personnel at the Client's cost and the Client shall issue a Variation to the Services in accordance with Clause 5.1 [*Variations*].

2.7 Client's Representative

2.7.1 The Client shall notify the Consultant of the extent of powers and authority delegated to the Client's Representative.

2.8 Services of Others

2.8.1 The Client shall arrange, at its own cost, for the provision of services from others as described in Appendix 2 [*Personnel, Equipment, Facilities and Services of Others to be Provided by the Client*], and the Consultant shall co-operate with the suppliers of such services but shall not be responsible for them or their performance. Where the provision of services by others is necessary for the proper performance of the Services then the Client shall ensure that such provision of services by others is given in sufficient time so as to allow the Consultant to proceed in accordance with the Programme.

2.8.2 Unless otherwise stated in Appendix 1 [*Scope of Services*], the responsibility for interface management between the Services and services provided by others remains with the Client.

2.4			
客户的资金安排		2.4.1	**客户**应在收到**咨询工程师（单位）**的任何请求后二十八（28）天内向**咨询工程师（单位）**提交合理的证据，证明已做出并正在保持资金安排，以使**客户**能够根据**附录3**[*报酬和付款*]或**协议书**的任何其他规定及时付款。
		2.4.2	如果**客户**打算对其资金安排进行任何重大改变，则应向**咨询工程师（单位）**发出**通知**，并附详细证明资料⊖。如果合理行事的**咨询工程师（单位）**对**客户**提交改变的建议和/或证明资料不满意，则**咨询工程师（单位）**有权根据第6.1.2项[*暂停服务*]（c）段的规定暂停**服务**。
2.5			
客户设备和设施的提供		2.5.1	**客户**出于**服务**的目的应向**咨询工程师（单位）**提供**附录2**[*由客户提供的人员、设备、设施和其他方的服务*]中所述的设备和设施，并适当考虑**进度计划**，且不收取费用。
2.6			
客户人员的提供		2.6.1	经与**咨询工程师（单位）**协商，**客户**应根据**附录2**[*由客户提供的人员、设备、设施和其他方的服务*]中的要求，如果有，为**咨询工程师（单位）**自费安排其雇用的适当合格人员的选择和提供。在提供**服务**时，根据适用法律的要求，此类人员只能接受**咨询工程师（单位）**的指示。
		2.6.2	**客户**提供的人员以及将来可能需要的任何替代人员，均须经**咨询工程师（单位）**认可。不得无理拒绝此类认可。
		2.6.3	如果**客户**不能提供其负责的**客户人员**，或者**咨询工程师（单位）**合理认为，**客户**提供给**咨询工程师（单位）**的人员的绩效不足以履行分配给他们的**服务**，**咨询工程师（单位）**则应安排另一批替代人员，费用由**客户**承担，**客户**应根据第5.1款[*变更*]的规定签发**服务变更**。
2.7			
客户代表		2.7.1	**客户**应告知**咨询工程师（单位）**付托给**客户代表**的权利范围和授权事项。
2.8			
其他方的服务		2.8.1	**客户**应自费安排**附录2**[*由客户提供的人员、设备、设施和其他方的服务*]所述的其他方提供的服务，**咨询工程师（单位）**应与此类**服务**的供应商合作，但不对其或其工作表现负责。如果其他方提供的服务是适当履行**服务**所必需的，则**客户**应确保其他方有足够的时间提供此类服务，以便**咨询工程师（单位）**能够按照**进度计划**进行。
		2.8.2	除非**附录1**[*服务范围*]中另有规定，否则**服务**与其他方提供的服务之间的接口管理责任仍由**客户**承担。

⊖ 此处中文按照勘误表改正后的英文翻译。——译者注

3 The Consultant

3.1 Scope of Services

3.1.1 The Consultant shall perform the Services as stated in Appendix 1 [*Scope of Services*].

3.1.2 The Consultant shall perform the Services in accordance with the Programme as may be amended from time to time in accordance with the Agreement.

3.1.3 The Consultant declares that as at the date of signature of the Agreement there are no circumstances or matters that may give rise to a conflict of interest in the performance of its obligations under the Agreement. The Consultant shall inform the Client immediately if it becomes aware of any such circumstances or matters. If a conflict of interest arises then the Parties shall agree, in good faith, on measures to manage such conflict.

3.2 Function and Purpose of Services

3.2.1 Where appropriate, the Client shall describe the function and purpose of the Services and state the same explicitly in Appendix 1 [*Scope of Services*].

3.3 Standard of Care

3.3.1 Notwithstanding any term or condition to the contrary in the Agreement or any related document or any legal requirement of the Country or any other relevant jurisdiction (including, for the avoidance of doubt, the jurisdiction of the place of establishment of the Consultant), in the performance of the Services the Consultant shall have no other responsibility than to exercise the reasonable skill, care and diligence to be expected from a consultant experienced in the provision of such services for projects of similar size, nature and complexity.

3.3.2 To the extent achievable using the standard of care in Sub-Clause 3.3.1, and without extending the obligation of the Consultant beyond that required under Sub-Clause 3.3.1, the Consultant shall perform the Services with a view to satisfying any function and purpose that may be described in Appendix 1 [*Scope of Services*].

3.3.3 The Consultant shall comply with all regulations, statutes, ordinances and other forms of standards, codes of practice and legislation applicable to the Services and the Agreement.

3.4 Client's Property

3.4.1 Anything supplied by or paid for by the Client for the use of the Consultant shall be the property of the Client and, where practicable, shall be so marked. The Consultant shall make reasonable endeavours to safeguard and protect such property of the Client until completion of the Services and/or return of such property to the Client.

3.5 Consultant's Personnel

3.5.1 The key personnel who are proposed by the Consultant to work in the Country shall be subject to acceptance by the Client with regard to their qualifications and experience. Such acceptance by the Client shall not be unreasonably withheld. Personnel, if any, included in the Consultant's offer/proposal included as part of the Agreement shall be deemed to be accepted by the Client on entering into the Agreement.

3 咨询工程师（单位）

3.1
服务范围

3.1.1　咨询工程师（单位）应履行附录1［*服务范围*］中规定的服务。

3.1.2　咨询工程师（单位）应履行根据协议书不时修订的进度计划提供的服务。

3.1.3　咨询工程师（单位）声明，截至协议书签署之日，在履行其协议书规定的义务时，不存在任何可能引起利益冲突的情况或事项。咨询工程师（单位）如发现任何此类情况或事项，应立即通知客户。如果发生利益冲突，双方应本着诚信原则，就管理此类冲突的措施达成商定。

3.2
服务的功能和目的

3.2.1　在适当情况下，客户应描述服务的功能和目的，并在附录1［*服务范围*］中明确说明。

3.3
谨慎标准

3.3.1　尽管协议书中或任何相关文件或工程所在国或任何其他相关司法管辖区［为避免疑问，包括咨询工程师（单位）成立地的司法管辖区］中有任何相反条款或条件，或任何法律要求，在履行服务的过程中，咨询工程师（单位）除了在类似规模、性质和复杂程度的项目中在提供此类服务方面所期望的经验丰富的咨询工程师（单位）运用合理的技能、谨慎和尽职之外，没有其他责任。

3.3.2　在使用第3.3.1项规定的谨慎标准可实现的范围内，在不将咨询工程师（单位）的义务扩展到第3.3.1项要求的范围之外的情况下，咨询工程师（单位）应履行服务，以满足附录1［*服务范围*］中所述的任何功能和目的。

3.3.3　咨询工程师（单位）应遵守适用于服务和协议书的所有法规、法令、条例和其他形式的标准、实践准则和法律。

3.4
客户财产

3.4.1　客户提供或支付给咨询工程师（单位）使用的任何物品应为客户的财产，在切实可行的情况下，应予以标注。咨询工程师（单位）应尽合理努力保障和保护客户的此类财产，直到服务完成和/或将此类财产归还给客户。

3.5
咨询工程师（单位）人员

3.5.1　咨询工程师（单位）建议的拟在工程所在国工作的关键人员的资格和经验应得到客户的认可。客户不得无理拒绝此类认可。作为协议书一部分的咨询工程师（单位）报价/建议书中包含的人员，如果有，应视为在签订协议书时已被客户接受。

3.6 Consultant's Representative

3.6.1 The Consultant shall notify the Client of the extent of powers and authority delegated to the Consultant's Representative.

3.6.2 If required by the Client, the Consultant shall designate an individual to liaise with the Client's Representative in the Country.

3.7 Changes in Consultant's Personnel

3.7.1 If it is necessary for any reason to replace any of the personnel provided by the Consultant, the Consultant shall arrange for replacement by a person(s) of suitable qualification and experience in the provision of the Services as soon as reasonably possible.

3.7.2 The cost of such replacement shall be borne by the Consultant except where the replacement is requested by the Client, and in such case:

(a) the request by the Client shall be made by Notice stating the reasons for it; such reasons shall relate to the provision of the Services and shall be reasonable and not vexatious; and

(b) the Client shall bear the cost of replacement unless misconduct or inability to perform satisfactorily in accordance with Sub-Clause 3.3.1 [*Standard of Care*] is the reason for the replacement of the relevant personnel by the Consultant.

3.8 Safety and Security of Consultant's Personnel

3.8.1 If in the reasonable opinion of the Consultant the health, safety or security of its personnel whilst in the Country is compromised by an Exceptional Event then the Consultant shall be entitled to suspend all or part of the Services in accordance with Sub-Clause 6.1.2 (b) [*Suspension of Services*] and remove such personnel from the Country until such time as the Exceptional Event has ceased.

3.9 Construction Administration

3.9.1 This Clause only applies where stated in the Particular Conditions and in Appendix 1 [*Scope of Services*], whereby the Consultant is required to perform the defined function of the engineer, employer's representative, project manager or similar under a Works Contract. Where such services are included in the scope of Services in Appendix 1 they shall be considered to be part of the Services.

3.9.2 Where explicitly described in Appendix 1 [*Scope of Services*], the Consultant shall perform the role of the engineer, employer's representative, project manager or similar as laid down in the Works Contract. The Consultant shall provide such construction administration services in accordance with the scope of Services.

3.9.3 When acting as the engineer, employer's representative, project manager or similar, the Consultant shall have the authority to act on behalf of the Client to the extent provided in the Works Contract. If the authority of the Consultant under the Works Contract is subject to prior approval of the Client, then the Client warrants that such restriction on the authority of the Consultant shall be stated in the Works Contract or shall be made known in writing to the contractor under the Works Contract. If the Consultant is authorised under the Works Contract to certify, determine or exercise discretion in the discharge of its duties then the Consultant shall act fairly as go-between the Client and the contractor, exercising independent professional judgement and using reasonable skill, care and diligence.

3.6 咨询工程师（单位）代表	3.6.1	咨询工程师（单位）应将付托给咨询工程师（单位）代表的权利和权限范围通知客户。
	3.6.2	如果客户要求，咨询工程师（单位）应指定一人与工程所在国的客户代表联络。
3.7 咨询工程师（单位）人员变动	3.7.1	如因任何原因有必要替代咨询工程师（单位）提供的任何人员，咨询工程师（单位）应尽快安排一名具有适当资格和经验的替代人员提供服务。
	3.7.2	此类替代的费用应由咨询工程师（单位）承担，客户要求替代的情况除外，在这种情况下：

（a） 客户的请求应以通知的形式提出，并说明理由；此类理由应与提供服务有关，且应合理而非无理取闹；（以及）

（b） 客户应承担替换的费用，不当行为或未能按照第3.3.1项[谨慎标准]的规定令人满意地履行职责是咨询工程师（单位）更换相关人员的原因的情况除外。

3.8 咨询工程师（单位）人员的安全和保障	3.8.1	如果咨询工程师（单位）合理认为其人员在工程所在国期间的健康、安全或保障会因例外事件受到损害，则咨询工程师（单位）有权根据第6.1.2项[服务暂停]（b）段的规定暂停全部或部分服务，并将这些人员从工程所在国撤出，直至例外事件停止。
3.9 施工管理	3.9.1	本条仅适用于专用条件和附录1[服务范围]中规定的情况，其中要求咨询工程师（单位）履行工程合同规定的工程师、雇主代表、项目经理或类似人员的规定职能。如果此类服务包含在附录1的服务范围内，则应将其视为服务的一部分。
	3.9.2	如果附录1[服务范围]中有明确规定，咨询工程师（单位）应履行工程合同中规定的工程师、雇主代表、项目经理或类似人员的职责。咨询工程师（单位）应根据服务范围提供此类施工管理服务。
	3.9.3	作为工程师、雇主代表、项目经理或类似人员行事时，咨询工程师（单位）应有权在工程合同规定的范围内代表客户行事。如果工程合同规定咨询工程师（单位）的权利须经客户事先批准，则客户保证，此类对咨询工程师（单位）权利的限制应在工程合同中加以明确，或应根据工程合同规定以书面形式告知承包商。如果工程合同授权咨询工程师（单位）在履行任务时确认、确定或行使自由裁量权，则咨询工程师（单位）应在客户和承包商之间公平行事，行使独立的专业判断，并运用合理的技能、谨慎和尽职。

3.9.4 The Consultant shall not be liable to the Client for the performance of the Works Contract by the contractor. In the discharge of its duties under the Works Contract, the Consultant shall only be liable to the Client if the Consultant commits a breach of the Agreement. In so far as the applicable law permits the Client shall indemnify the Consultant against any and all claims made by the contractor against the Consultant arising out of or connected with the Works Contract.

3.9.5 The Consultant shall not be liable to the Client or the contractor for the means, techniques, methods or sequencing of any aspect of the Works Contract or for the safety or adequacy of any of the contractor's operations.

3.9.6 If an ambiguity or discrepancy is found between the Consultant's obligations under the Agreement and the Consultant's duties under the Works Contract, the Consultant shall give Notice to the Client indicating the effect of such ambiguity or discrepancy. The Client shall rectify such ambiguity or discrepancy by instruction as soon as reasonably practicable and where necessary shall issue a Variation to the Services in accordance with Clause 5.1 [*Variations*].

4 Commencement and Completion

4.1 Agreement Effective

4.1.1 The Agreement shall be effective from the date of the latest signature necessary to complete the formal Agreement (the "Effective Date").

4.2 Commencement and Completion of Services

4.2.1 The Consultant shall commence the performance of the Services as soon as is reasonably practicable after the Commencement Date. The Consultant shall complete the whole of the Services within the Time for Completion.

4.3 Programme

4.3.1 Within fourteen (14) days of the Commencement Date the Consultant shall submit its Programme which shall include as a minimum:

a) the order and timing in which the Consultant intends to carry out the Services in order to complete the Services within the Time for Completion;

b) any key dates stipulated in Appendix 4 [*Programme*] or elsewhere in the Agreement for the delivery of any part of the Services to the Client;

c) the key dates when decisions, consents, approvals or information from the Client or third parties is required to be given to the Consultant;

d) any other requirements stated in Appendix 4 [*Programme*].

The Consultant shall keep the Programme under review and shall amend the same as and when necessary to comply with the Agreement.

4.3.2 Unless the Client, within fourteen (14) days of receiving the Programme, gives Notice to the Consultant stating the extent to which it does not comply with the Agreement, the Consultant shall proceed in accordance with the Programme, subject to its other obligations under the Agreement.

3.9.4 **咨询工程师（单位）**就承包商履行工程合同不对**客户**负责。在履行其工程合同规定的任务时，只有在**咨询工程师（单位）**违反**协议书**的情况下，**咨询工程师（单位）**才应对**客户**负责。在适用法律允许的范围内，**客户**应保障**咨询工程师（单位）**免受承包商因工程合同而向**咨询工程师（单位）**提出的任何及所有索赔。

3.9.5 对于工程合同的任何手段、技术、方法或顺序，或承包商任何操作的安全性或充分性，**咨询工程师（单位）**均不对**客户**或承包商负责。

3.9.6 如果发现**协议书**规定的**咨询工程师（单位）**义务与工程合同规定的**咨询工程师（单位）**任务之间存在歧义或不一致，**咨询工程师（单位）**应**通知客户**，说明此类歧义或不一致的影响。**客户**应在合理可行的情况下尽快通过指示纠正此类歧义或不一致，并在必要时根据**第5.1款[变更]**的规定签发**服务**变更。

4 开始和完成

4.1 协议书生效

4.1.1 **协议书**应自完成正式**协议书**所需的最新签字之日起生效（"生效日期"）。

4.2 服务的开始和完成

4.2.1 **咨询工程师（单位）**应在**开始日期**后合理可行的情况下尽快开始履行**服务**。**咨询工程师（单位）**应在**完成时间**内完成全部**服务**。

4.3 进度计划

4.3.1 **咨询工程师（单位）**应在**开始日期**后十四（14）天内提交其**进度计划**，应至少包括：

a) **咨询工程师（单位）**为在**完成时间**内完成**服务**打算实施**服务**的顺序和时间；

b) **附录4[进度计划]**或**协议书**另外规定的向**客户**交付任何部分**服务**的关键日期；

c) 需要向**咨询工程师（单位）**提供的**客户**或第三方的决定、同意、批准或信息交流的关键日期；

d) **附录4[进度计划]**中规定的任何其他要求。

咨询工程师（单位）应始终保持审核**进度计划**，并应在必要时对其进行修改，以遵守**协议书**。

4.3.2 除非**客户**在收到**进度计划**后十四（14）天内向**咨询工程师（单位）**发出**通知**，说明其不遵守**协议书**的程度，否则**咨询工程师（单位）**应在遵守**协议书**规定的其他义务的前提下，按照**进度计划**继续进行。

4.3.3   The Parties shall promptly give Notice to each other of any specific, actual or probable future events or circumstances which may adversely affect or delay the Services or lead to an increase in the cost of the Services.

4.4 Delays

4.4.1   The Consultant shall be entitled to an extension of the Time for Completion if and to the extent that completion of the Services is or will be delayed by any of the following causes:

a)   a Variation to the Services;
b)   any delay, impediment or prevention caused by or attributable to the Client, or the Client's other consultants, contractors, or other third parties;
c)   an Exceptional Event; or
d)   any other event or circumstance giving an entitlement to extension of the Time for Completion under the Agreement.

4.4.2   Any extension of the Time for Completion shall have due regard to the Programme and any constraints therein.

4.4.3   Where any circumstance referred to in Sub-Clause 4.4.1 causes the Consultant to incur Exceptional Costs, then the agreed remuneration shall be adjusted in accordance with Sub-Clause 7.1.2 [*Payment to the Consultant*]. As soon as reasonably practicable the Consultant shall inform the Client of the occurrence of the Exceptional Costs by issue of a Notice.

4.5 Rate of Progress of Services

4.5.1   If, for any reason that does not entitle the Consultant to an extension of the Time for Completion, the rate of progress of the Services is, in the reasonable opinion of the Client, insufficient to ensure completion of the Services within the Time for Completion, then the Client may give Notice to that effect to the Consultant. Upon receipt of such Notice the Consultant shall revise the Programme and shall issue a Notice to the Client describing the measures the Consultant intends to put in place in order to complete the Services in accordance with the Time for Completion.

4.6 Exceptional Event

4.6.1   If a Party is prevented from performing any of its obligations under the Agreement by, or due to, an Exceptional Event then it shall give a Notice to the other Party providing a description of the Exceptional Event together with an assessment of its effects on the Party's ability to comply with its obligations under the Agreement. The Notice shall be given within fourteen (14) days from when the Party becomes aware, or should have become aware, of the event or circumstance constituting an Exceptional Event.
The Party having given Notice, shall be excused from performance of such obligations for so long as the effects of the Exceptional Event prevent such performance.

4.6.2   Where an Exceptional Event gives rise to an unavoidable change in the scope of Services then the Client shall issue a Variation to the Services in accordance with Clause 5.1 [*Variations*]. Where an Exceptional Event gives rise to a delay in the completion of the Services then the Consultant shall be entitled to an extension of the Time for Completion in accordance with Clause 4.4 [*Delays*]

| | | 4.3.3 | 任何可能对**服务**产生不利影响或延误，或导致**服务**成本增加的任何具体、实际或可能的未来事件或情况，双方应立即**通知**对方。 |

4.4 延误

| | 4.4.1 | 如因以下任何原因而延迟或将延迟**服务**的完成，**咨询工程师（单位）**有权延长**完成时间**： |

a) **服务变更**；
b) 由**客户**或**客户**的其他咨询工程师、承包商或其他第三方造成的或可归因于其的任何延误、阻碍或阻止；

c) **例外事件**；（或）
d) 根据**协议书**，有权延长**完成时间**的任何其他事件或情况。

4.4.2 **完成时间**的任何延长应适当考虑**进度计划**及其任何限制。

4.4.3 如果第 4.4.1 项所述的任何情况导致**咨询工程师（单位）**产生**例外费用**，则商定的报酬应根据第 7.1.2 项 [*对咨询工程师（单位）的付款*] 进行调整。在合理可行的情况下，**咨询工程师（单位）**应尽快签发**通知**将**例外费用**的发生情况告知**客户**。

4.5 服务进度

4.5.1 如果由于**咨询工程师（单位）**无权延长**完成时间**的原因，**客户**合理认为**服务**进度不足以确保在**完成时间**内完成**服务**，则**客户**可就此向**咨询工程师（单位）**发出**通知**。收到此类通知后，**咨询工程师（单位）**应修改**进度计划**，并应向**客户**签发**通知**，说明**咨询工程师（单位）**为按照**完成时间**完成**服务**而打算采取的措施。

4.6 例外事件

4.6.1 如果一方被或因**例外事件**阻止而无法履行其**协议书**规定的任何义务，则该方应向另一方发出**通知**，描述该**例外事件**并评估其对该方履行**协议书**规定的义务的能力的影响。该**通知**应在该方意识到或应已意识到构成**例外事件**的事件或情况后的十四（14）天内发出。

只要**例外事件**的影响阻止了义务的履行，就应免除已发出**通知**的一方履行此类义务。

4.6.2 如果**例外事件**导致**服务**范围不可避免地发生变化，则**客户**应根据第 5.1 款 [*变更*] 的规定签发**服务变更**。如果**例外事件**导致**服务**完成的延误，**咨询工程师（单位）**则有权根据第 4.4 款 [*延误*] 的规定延长**完成时间**。

4.6.3 Notwithstanding any other provision of this Clause 4.6, the obligations of either Party to make payments to the other Party under the Agreement shall not be excused by an Exceptional Event.

5 Variations to Services

5.1 Variations

5.1.1 A Variation to the Services may be initiated by the Client by issue of a Variation Notice at any time prior to completion of the Services. The Client may request the Consultant to submit a proposal in respect of a proposed Variation. If the proposal is accepted by the Client then the Variation shall be confirmed by the Client by issue of a Variation Notice. Any such Variation shall not substantially change the extent or nature of the Services.

5.1.2 A Variation to the Services may be issued in respect of any:

(a) amendment to Appendix 1 [*Scope of Services*] or to Appendix 2 [*Personnel, Equipment, Facilities and Services of Others to be Provided by the Client*];

(b) omission of part of the Services but only where such omitted services are no longer required by the Client;

(c) changes in the specified sequence or timing of the performance of the Services;

(d) changes in the method of implementation of the Services;

(e) provision of the Agreement requiring the issue of a Variation; or

(f) proposal submitted by the Consultant (at the Client's request or otherwise) and accepted in writing by the Client.

5.1.3 The Consultant shall give Notice to the Client as soon as reasonably practicable where the Consultant considers that any instruction or direction from the Client or any other circumstance constitutes a Variation to the Services. The Consultant shall include in the Notice details of the estimated impact upon the Programme and cost of the Services for such matters. Within fourteen (14) days of receipt of the Notice the Client shall either issue a Variation Notice, or cancel the instruction or direction, or state by issue of a further Notice why the Client considers the instruction, direction or circumstance does not constitute a Variation to the Services. In such case the Consultant shall comply with and be bound by such further Notice unless the Consultant refers the matter as a dispute under Clause 10 [*Disputes and Arbitration*] within seven (7) days of receipt of such further Notice.

5.1.4 Unless the Consultant promptly gives Notice to the Client (with supporting evidence) that:

(a) it does not possess the relevant skills or resources to carry out the Variation; or

(b) the Consultant considers that the Variation will substantially change the extent or nature of the Services;

the Consultant shall be bound by each Variation. The Consultant shall not otherwise make any changes to the Services.

4.6.3　尽管**第 4.6 款**有任何其他规定，任一方根据**协议书**向另一方付款的义务不得因**例外事件**而免除。

---

5 服务变更

**5.1
变更**

5.1.1　**客户**可在**服务**完成前的任何时间签发**变更通知**，对**服务**进行**变更**。**客户**可要求**咨询工程师（单位）**就拟议的**变更**提交建议书。如果建议书被**客户**接受，则**客户**应通过签发**变更通知**确认**变更**。任何此类**变更**不得实质性地改变**服务**的范围或性质。

5.1.2　**服务变更**可就以下任何事项签发：

（a）　对附录 1 [*服务范围*] 或附录 2 [*由客户提供的人员、设备、设施和其他方的服务*] 的修订；

（b）　部分**服务**的删减，但仅限于**客户**不再需要此类删减的服务时；

（c）　履行**服务**的特定顺序或时间安排发生改变；

（d）　**服务**实施方法的改变；

（e）　要求签发**变更**的**协议书**规定；（或）

（f）　由**咨询工程师（单位）**提交的（应**客户**要求或其他方式）并由**客户**书面接受的建议书。

5.1.3　如果**咨询工程师（单位）**认为**客户**的任何指示或指导或任何其他情况构成对**服务**的变更，**咨询工程师（单位）**应在合理可行的情况下尽快**通知客户**。**咨询工程师（单位）**应在**通知**中详细说明此类事项对**进度计划**和**服务**成本的影响。在收到**通知**后十四（14）天内，**客户**应签发**变更通知**，或取消指示或指导，或签发进一步**通知**，说明**客户**认为该指示、指导或情况不构成对**服务**的**变更**的原因。在这种情况下，**咨询工程师（单位）**应遵守该进一步**通知**并受其约束，除非**咨询工程师（单位）**在收到该进一步**通知**后的七（7）天内，根据**第 10 条** [*争端和仲裁*] 的规定将该事项作为争端提交。

5.1.4　除非**咨询工程师（单位）**及时**通知客户**（附支持证据），说明：

（a）　其不具备执行**变更**的相关技能或资源；（或）

（b）　**咨询工程师（单位）**认为**变更**将实质性地改变**服务**的范围或性质；

咨询工程师（单位）应受每项**变更**的约束。**咨询工程师（单位）**不得以其他方式对**服务**进行任何**改变**。

5.2
Agreement of Variation Value and Impact

5.2.1 The Client and the Consultant shall agree the value of any Variation, or its method of calculation, including its impact (if any) upon other parts of the Services, the Programme and the Time for Completion.

5.2.2 The value of any Variation shall be determined in accordance with or based upon the rates and/or prices in Appendix 3 [*Remuneration and Payment*]. Where the rates and/or prices are not applicable to the Variation then new rates shall be agreed by the Parties.

5.2.3 The value of the Variation and its impact on the Programme shall be agreed and confirmed in writing by the Client to the Consultant. Pursuant to such agreement the Client shall issue an instruction to the Consultant to commence work on the Variation.

5.2.4 Where agreement under Sub-Clause 5.2.3 is not reached within fourteen (14) days of receipt by the Consultant of the Variation Notice or it is not practicable to establish and agree between the Parties all the effects of the Variation prior to the Consultant commencing work on the Variation then the Client may by Notice instruct the Consultant to commence work on the Variation and the Consultant shall comply with such instruction. The Consultant shall be compensated on a time spent basis at the rates and prices stated in Appendix 3 [*Remuneration and Payment*] or if no rates and prices are stated then at reasonable rates and prices until such time as agreement is reached on all the effects of the Variation.

6 Suspension of Services and Termination of Agreement

6.1
Suspension of Services

6.1.1 The Client may suspend all or part of the Services at its sole discretion and for any reason by giving twenty-eight (28) days' Notice to the Consultant.

6.1.2 The Consultant may suspend all or part of the Services in the following circumstances:

a) When the Consultant has not received payment of an invoice or a part of an invoice, as the case may be, by the due date for payment of such invoice and the Client has not issued a valid Notice in accordance with Clause 7.5 [*Disputed Invoices*] stating the reasons for non-payment of the invoice or part thereof, subject to the Consultant giving seven (7) days' Notice to the Client.

b) Where an Exceptional Event arises, including that contemplated under Clause 3.8 [*Safety and Security of Consultant's Personnel*]. Notice shall be given to the Client as soon as reasonably practicable. The Consultant shall take reasonable endeavours to avoid or minimise such suspension of all or part of the Services.

c) Failure by the Client to satisfy the requirements of Clause 2.4 [*Client's Financial Arrangements*].

6.2
Resumption of Suspended Services

6.2.1 When the Services have been suspended under Sub-Clause 6.1.1 [*Suspension of Services*] the Consultant shall resume the Services or part thereof, as the case may be, within twenty-eight (28) days' of receipt of Notice from the Client instructing the Consultant to resume the Services or part thereof.

5.2		
变更价值和影响协议书	5.2.1	**客户**和**咨询工程师（单位）**应商定任何**变更**的价值或其计算方法，包括其对**服务**其他部分、**进度计划**和**完成时间**的影响（如果有）。
	5.2.2	任何**变更**的价值应根据或基于**附录3**［*报酬和付款*］中的费率和/或价格确定。如果费率和/或价格不适用于**变更**，则双方应商定新的费率。
	5.2.3	**变更**的价值及其对**进度计划**的影响应由**客户**以书面形式同意，并向**咨询工程师（单位）**确认。根据此类商定，**客户**应向**咨询工程师（单位）**签发指示，开始**变更**工作。
	5.2.4	如果**咨询工程师（单位）**在收到**变更通知**后的十四（14）天内未能根据第**5.2.3**项达成商定，或者在**咨询工程师（单位）**开始**变更**工作之前，无法在双方之间确定和商定**变更**的所有影响，则**客户**可通过**通知**指示**咨询工程师（单位）**开始**变更**工作，**咨询工程师（单位）**应遵守此类指示。**咨询工程师（单位）**应按照**附录3**［*报酬和付款*］中规定的费率和价格基于所花费的时间得到补偿，如未规定费率和价格，则应以合理的费率和价格进行补偿，直到就**变更**的所有影响达成商定。

6 服务暂停和协议书终止

6.1		
服务暂停	6.1.1	**客户**可出于任何原因，可提前二十八（28）天通知**咨询工程师（单位）**，自行决定暂停全部或部分**服务**。
	6.1.2	在下列情况下，**咨询工程师（单位）**可暂停全部或部分**服务**：
	a)	当**咨询工程师（单位）**在该发票的付款到期日前未收到发票或部分发票的付款，视情况而定，且**客户**未能根据第**7.5款**［*有争议发票*］签发有效**通知**，说明未支付发票或部分发票的原因时，以**咨询工程师（单位）**提前七（7）天**通知客户**为准。
	b)	发生**例外事件**时，包括第**3.8**款［*咨询工程师（单位）人员的安全和保障*］规定的事件。应在合理可行的情况下尽快**通知客户**。**咨询工程师（单位）**应尽合理努力避免或尽量减少全部或部分**服务**的暂停。
	c)	在**咨询工程师（单位）**提前七（7）天通知**客户**的情况下，**客户**未能满足第**2.4**款［*客户的资金安排*］的要求⊖。

6.2		
服务暂停的恢复	6.2.1	根据第**6.1.1**项［*服务暂停*］的规定暂停**服务**时，**咨询工程师（单位）**应在收到**客户**指示**咨询工程师（单位）**恢复**服务**或其部分**服务**的**通知**后二十八（28）天内，恢复**服务**或部分**服务**，视情况而定。

⊖ 此处中文按照勘误表改正后的英文翻译。——译者注

6.2.2 Where the Services have been suspended under Sub-Clause 6.1.2 [*Suspension of Services*] the Consultant shall resume the Services or part thereof, as the case may be, as soon as reasonably practicable after the matters giving rise to the suspension have ceased.

6.3 Effects of Suspension of the Services

6.3.1 The Consultant shall be paid for Services performed in accordance with the Agreement up to the date of suspension of the Services or part thereof, as the case may be.

6.3.2 During the period of suspension the Consultant shall not perform the Services or part thereof as the case may be, but shall ensure, so far as is reasonably practicable, the security, maintenance and custody of the Services so as to prevent spoilage or loss.

6.3.3 If during the suspension and resumption of Services or part thereof the Consultant incurs Exceptional Costs, then:

(a) the agreed remuneration shall be adjusted in accordance with Sub-Clause 7.1.2 [*Payment to the Consultant*];
(b) the Time for Completion shall be amended in accordance with Clause 4.4 [*Delays*] to reflect the effect of the suspension on the Programme.
(c) as soon as reasonably practicable the Consultant shall inform the Client by issue of a Notice of the occurrence of these Exceptional Costs.

6.3.4 The Consultant shall take reasonable measures to mitigate the effects of the suspension of the Services or part thereof.

6.4 Termination of Agreement

6.4.1 Termination by the Client

(a) If the Consultant without good reason is in breach of a material term or condition of the Agreement, the Client may give Notice to the Consultant outlining the breach and the remedy required under the Agreement. If the Consultant has not proceeded to remedy the breach within twenty-eight (28) days after the issue of the Notice then the Client may terminate the Agreement upon giving fourteen (14) days' Notice to the Consultant.

(b) Notwithstanding the notice periods in Sub-Clause 6.4.1 (a), if the Consultant becomes bankrupt or insolvent, goes into liquidation, has a receiving or administration order made against it, compounds with its creditors, or carries on business under a receiver, trustee or manager for the benefit of its creditors, or if any act is done or event occurs which (under applicable laws) has a similar effect to any of these acts or events, the Client may in so far as the applicable laws permit terminate the Agreement with immediate effect upon service of an appropriate Notice.

(c) Notwithstanding the notice periods in Sub-Clause 6.4.1 (a), if the Consultant is in breach of Clause 1.10 [*Anti - Corruption*], the Client may terminate the Agreement with immediate effect upon service of an appropriate Notice.

(d) At its sole discretion upon giving the Consultant fifty-six (56) days' Notice provided always that the Client shall not be entitled to use this provision in order to obtain the Services from others, or in order to perform the Services by itself.

6.2.2 如果根据第 6.1.2 项 [*服务暂停*] 的规定已暂停**服务**，**咨询工程师（单位）**应在导致暂停的事项停止后，在合理可行的范围内尽快恢复**服务**或其部分**服务**，视情况而定。

6.3 服务暂停的影响

6.3.1 **咨询工程师（单位）**应根据**协议书**履行的**服务**获得付款，直至**服务**或部分**服务**暂停之日，视情况而定。

6.3.2 暂停期间，**咨询工程师（单位）**不得履行**服务**或部分**服务**，视情况而定，但应在合理可行的情况下，确保**服务**的安保、维护和保管，以防止损坏或损失。

6.3.3 在暂停和恢复**服务**或部分**服务**期间，如果**咨询工程师（单位）**招致了**例外费用**，则：

(a) **咨询工程师（单位）**应在合理可行的情况下，尽快签发**通知**将这些**例外费用**的发生情况告知**客户**；（以及）

(b) 商定的报酬应根据第 7.1.2 项 [*对咨询工程师（单位）的付款*] 的规定进行调整。⊖

6.3.4 完成时间应根据第 4.4 款 [*延误*] 的规定进行修改，以反映暂停对进度计划的影响。⊖

6.3.5 **咨询工程师（单位）**应采取合理措施减轻暂停**服务**或部分**服务**的影响。⊖

6.4 协议书终止

6.4.1 由**客户**终止

(a) 如果**咨询工程师（单位）**在无正当理由的情况下违反**协议书**的重要条款或条件，**客户**可向**咨询工程师（单位）**发出**通知**，说明违约情况和**协议书**要求的补救措施。如果**咨询工程师（单位）**在签发**通知**后二十八（28）天内未采取补救措施，则**客户**可在提前十四（14）天**通知咨询工程师（单位）**后终止**协议书**。

(b) 尽管有第 6.4.1 项（a）段规定的通知期限，如果**咨询工程师（单位）**破产或资不抵债、进行清算、收到对其发出的接管令或管理令、与其债权人达成和解，或在接管人、受托人或管理人的监督下为其债权人的利益开展业务，或者，如果任何行为或发生的事件（根据适用法律）与上述任何行为或事件具有类似的影响，**客户**可在适用法律允许的范围内，在发出适当**通知**后立即终止**协议书**。

(c) 尽管有第 6.4.1 项（a）段规定的通知期限，如果**咨询工程师（单位）**违反第 1.10 款 [*反腐败*] 的规定，**客户**可在送达适当**通知**后立即终止**协议书**。

(d) 尽管可自主决定向**咨询工程师（单位）**提前五十六（56）天发出**通知**，但**客户**应无权使用本规定以便从其他方获得**服务**，或自己履行**服务**。

⊖ 此处中文按照勘误表改正后的英文翻译。——译者注

(e) Without prejudice to Sub-Clause 6.1.1 [*Suspension of Services*], where an Exceptional Event has led to a suspension of the Services for more than one hundred and sixty-eight (168) days the Client may terminate the Agreement upon giving fourteen (14) days' Notice to the Consultant.

6.4.2 Termination by the Consultant

(a) If the Services have been suspended under Sub-Clause 6.1.1 [*Suspension of Services*] for more than one hundred and sixty-eight (168) days the Consultant may terminate the Agreement upon giving fourteen (14) days' Notice to the Client.

(b) If the Services have been suspended under Sub-Clause 6.1.2 (a) [*Suspension of Services*] or Sub-Clause 6.1.2 (c) [*Suspension of Services*] for more than forty-two (42) days the Consultant may terminate the Agreement upon giving fourteen (14) days' Notice to the Client.

(c) If the Client becomes bankrupt or insolvent, goes into liquidation, has a receiving or administration order made against it, compounds with its creditors, or carries on business under a receiver, trustee or manager for the benefit of its creditors, or if any act is done or event occurs which (under applicable laws) has a similar effect to any of these acts or events, the Consultant may in so far as the applicable laws permit terminate the Agreement with immediate effect upon service of an appropriate Notice.

(d) If the Client is in breach of Clause 1.10 [*Anti – Corruption*] the Consultant may terminate the Agreement with immediate effect upon service of an appropriate Notice.

(e) If the Services have been suspended under Sub-Clause 6.1.2 (b) [*Suspension of Services*] for more than one hundred and sixty-eight (168) days the Consultant may terminate the Agreement upon giving fourteen (14) days' Notice to the Client.

6.5 Effects of Termination

6.5.1 The Consultant shall be paid for Services performed in accordance with the Agreement up to the date of termination of the Agreement.

6.5.2 If the Agreement is terminated in accordance with Sub-Clause 6.4.1 (a) or (b) or (c) [*Termination of Agreement*] the Client shall, without prejudice to any other rights the Client may have under the Agreement, be entitled to:

(a) take over from the Consultant all documents, information, calculations and other deliverables, whether in electronic format or otherwise, pertaining to the Services performed up to the date of termination, necessary to enable the Client to complete the Services either by itself or with the assistance of another consultant (all documents in electronic format shall be editable);

(b) claim compensation for reasonable costs directly incurred as a consequence of the termination, including but not limited to additional costs incurred in arranging for the Services to be completed by another consultant;

(c) withhold payments due to the Consultant until all the costs incurred by the Client under Sub-Clause 6.5.2 (b) above have been established and all documents, information, calculations and other deliverables necessary to enable the Client to complete the Services have been received. The Client shall act expeditiously and without delay in establishing its own costs under Sub-clause 6.5.2 (b).

（e） 如果[注]**例外事件**导致**服务**暂停超过一百六十八（168）天，**客户**可在提前十四（14）天通知**咨询工程师（单位）**后终止**协议书**。

  6.4.2 由**咨询工程师（单位）**终止

（a） 如果根据第6.1.1项［*服务暂停*］的规定暂停**服务**超过一百六十八（168）天，**咨询工程师（单位）**可在提前十四（14）天通知**客户**后终止**协议书**；

（b） 如果根据第6.1.2项［*服务暂停*］（a）段或第6.1.2项［*服务暂停*］（c）段暂停**服务**超过四十二（42）天，**咨询工程师（单位）**可在提前十四（14）天通知**客户**后终止**协议书**；

（c） 如果**客户**破产或资不抵债、进行清算、收到针对其发出的接管令或管理令、与其债权人达成和解，或在接管人、受托人或管理人的监督下为其债权人的利益开展业务，或者，如果采取的任何行为或发生的事件（根据适用法律）与上述任何行为或事件具有类似的影响，**咨询工程师（单位）**可在适用法律允许的范围内，在发出适当**通知**后立即终止**协议书**；

（d） 如果**客户**违反第1.10款［*反腐败*］的规定，**咨询工程师（单位）**可在送达适当**通知**后立即终止**协议书**；

（e） 如果根据第6.1.2项［*服务暂停*］（b）段的规定暂停**服务**超过一百六十八（168）天，**咨询工程师（单位）**可在提前十四（14）天向**客户**发出**通知**后终止**协议书**。

| 6.5 终止的影响 | 6.5.1 | **咨询工程师（单位）**应在**协议书**终止日前根据**协议书**履行的**服务**获得报酬。 |

  6.5.2 如果**协议书**根据第6.4.1项［*协议书终止*］（a）段或（b）或（c）段的规定终止，在不损害**客户**可能享有的**协议书**规定的任何其他权利的情况下，**客户**有权：

（a） 从**咨询工程师（单位）**处接管截至终止日期与**服务**相关的、使**客户**能够自行或在其他咨询工程师的协助下完成**服务**所必需的所有文件、信息、计算和其他可交付成果，无论是电子格式还是其他格式（所有电子格式的文件均应可编辑）；

（b） 要求赔偿因终止而直接产生的合理费用，包括但不限于安排另一咨询工程师完成**服务**而产生的额外费用；

（c） 扣留应付给**咨询工程师（单位）**的款项，直至**客户**根据上述第6.5.2项（b）段承担的所有费用均已确定，且已收到使**客户**能够完成**服务**所需的所有文件、信息、计算和其他可交付成果。**客户**应尽快行动，根据第6.5.2款（b）段确定其的费用。

[注] 此处中文按照勘误表改正后的英文翻译。——译者注

The Client shall take all reasonable steps to mitigate such costs. The Client's entitlement under this Sub-Clause 6.5.2 shall be limited to those costs that are reasonably foreseeable at the time of signature of the Agreement.

6.5.3 If the Agreement is terminated in accordance with Sub-Clause 6.4.1 (d) or (e) or Sub-Clause 6.4.2 [*Termination of Agreement*] and the Consultant incurs Exceptional Costs, then, without prejudice to any other rights the Consultant may have under the Agreement, the agreed remuneration shall be adjusted in accordance with Sub-Clause 7.1.2 [*Payment to the Consultant*]. The Consultant shall inform the Client as soon as reasonably practicable by issue of a Notice of the occurrence of the Exceptional Costs.

6.5.4 Where the Agreement is terminated under Sub-Clause 6.4.1 (d) or Sub-Clause 6.4.2 (a) to (d) [*Termination of Agreement*] then the Consultant shall be entitled to be paid the loss of profit that would otherwise have been earned on the Services not performed due to the termination.

6.6 Rights and Liabilities of the Parties

6.6.1 Termination of the Agreement shall not prejudice or affect the accrued rights or claims and liabilities of the Parties.

7 Payment

7.1 Payment to the Consultant

7.1.1 The Client shall pay the Consultant for the Services (including Variations to the Services) in accordance with the details stated in Appendix 3 [*Remuneration and Payment*].

7.1.2 Unless otherwise agreed in writing, the Client shall pay the Consultant in respect of Exceptional Costs:

(a) for the extra time spent by the Consultant's personnel in the performance of the Services at the rates and prices stated in Appendix 3 [*Remuneration and Payment*]. Where the rates and prices are not applicable then new rates and prices shall be agreed by the Parties. If agreement is not reached within fourteen (14) days of the issue of the relevant Notice then reasonable rates and prices shall be applied; and

(b) the cost of all other expenses reasonably incurred by the Consultant.

7.1.3 The Client shall pay any other amounts that become due under the Agreement.

7.2 Time for Payment

7.2.1 Amounts due to the Consultant shall be paid within twenty-eight (28) days of the date of issue of the Consultant's invoice unless otherwise stated in Appendix 3 [*Remuneration and Payment*].

7.2.2 If the Consultant does not receive payment within the time stated in Sub-Clause 7.2.1 it shall be paid financing charges at the rate(s) stated in Appendix 3 [*Remuneration and Payment*] compounded monthly on the amount overdue and in its currency calculated from the due date for payment of the invoice to the actual date payment is received from the Client. Such financing charges shall not affect the rights of the Consultant stated in Sub-Clause 6.1.2 (a) [*Suspension of Services*] or Sub-Clause 6.4.2 [*Termination of Agreement*].

客户应采取一切合理措施来减少此类费用。第 6.5.2 项规定的客户权利应限于签署协议书时可合理预见的费用。

6.5.3　如果协议书根据第 6.4.1 项（d）段或（e）段或第 6.4.2 项［*协议书终止*］的规定终止，且咨询工程师（单位）产生了**例外费用**，则在不损害协议书规定的咨询工程师（单位）可能享有的任何其他权利的情况下，商定的报酬应根据第 7.1.2 项［*对咨询工程师（单位）付款*］的规定进行调整。咨询工程师（单位）应在合理可行的情况下，尽快向客户发出产生例外费用的通知。

6.5.4　如果协议书根据第 6.4.1 项（d）段或第 6.4.2 项（a）至（d）段［*协议书终止*］的规定终止，则咨询工程师（单位）应有权获得因终止而未履行的服务本应获得的利润损失。

6.6 双方权利和责任

6.6.1　协议书的终止不应损害或影响双方的应计权利或索赔和责任。

7 付款

7.1 对咨询工程师（单位）的付款

7.1.1　客户应按照附录 3［*报酬和付款*］中的详细规定，向咨询工程师（单位）支付服务费用（包括服务变更）。

7.1.2　除非另有书面商定，否则客户应向咨询工程师（单位）支付例外费用：

（a）咨询工程师（单位）人员根据附录 3［*报酬和付款*］中规定的费率和价格在履行服务过程中花费的额外时间。如果费率和价格不适用，则双方应商定新的费率和价格。如果在发出相关通知后十四（14）天内未能达成商定，则应采用合理的费率和价格；（以及）

（b）咨询工程师（单位）合理产生的所有其他支出的成本。

7.1.3　客户应支付协议书规定的应付的任何其他款项。

7.2 付款时间

7.2.1　除非附录 3［*报酬和付款*］另有规定，否则应在咨询工程师（单位）发票开具之日起二十八（28）天内向咨询工程师（单位）支付应付款额。

7.2.2　如果咨询工程师（单位）在第 7.2.1 项规定的时间内未收到付款，则应按照附录 3［*报酬和付款*］中规定的费率，就逾期金额按月复利和货币支付融资费用，并从发票付款到期日至从客户处收到付款的实际日期计算。此类融资费用不应影响第 6.1.2 项（a）段［*服务暂停*］或第 6.4.2 项［*协议书终止*］中规定的咨询工程师（单位）的权利。

7.2.3 Without prejudice to Sub-Clause 6.5.2 (c) [*Effects of Termination*] the Client shall not withhold payment of any part of an invoice for any amount properly due to the Consultant under the Agreement by reason of claims or alleged claims against the Consultant unless the amount to be withheld has been agreed with the Consultant as due to the Client, or has been awarded by an adjudicator or an arbitrator to the Client pursuant to a referral under Clause 10 [*Disputes and Arbitration*].

7.3 Currencies of Payment

7.3.1 The currencies applicable to the Agreement are those stated in Appendix 3 [*Remuneration and Payment*].

7.3.2 If at the Effective Date of the Agreement or during the performance of the Services the conditions in the Country (except where the Country is the principal place of business of the Consultant) are such as may:

(a) prevent or delay the transfer abroad of Local or Foreign Currency payments received by the Consultant in the Country;

(b) restrict the availability or use of Foreign Currency in the Country; or

(c) impose taxes or differential rates of exchange for the transfer from abroad of Foreign Currency into the Country by the Consultant for Local Currency expenditure and subsequent re-transfer abroad of Foreign Currency or Local Currency up to the same amount, such as to inhibit the Consultant in the performance of the Services or to result in financial disadvantage to it, then the Client agrees that such circumstances shall be deemed to justify the application of Clause 4.6 [*Exceptional Event*] if alternative financial arrangements are not made to the satisfaction of the Consultant.

7.4 Third-Party Charges on the Consultant

7.4.1 Except where specified in the Particular Conditions or Appendix 3 [*Remuneration and Payment*] and except where the Country is the principal place of business of the Consultant:

(a) the Client shall whenever possible arrange that exemption is granted to the Consultant and those of its personnel who are not normally resident in the Country from any payments required by the government or authorised third parties in the Country which arise from the Agreement in respect of:
 (i) their remuneration;
 (ii) their imported goods other than food and drink;
 (iii) goods imported for the Services;
 (iv) documents imported for the Services;

(b) whenever the Client is unsuccessful in arranging such exemption, it shall reimburse the Consultant for such payments properly made; provided that the goods or documents imported for the Services when no longer required for the purpose of the Services and not the property of the Client:
 (i) shall not be disposed of in the Country without the Client's approval;
 (ii) shall not be exported without payment to the Client of any refund or rebate recoverable and received from the government or authorised third parties.

7.2.3 在不影响第 6.5.2 项（c）段［*终止的影响*］的情况下，**客户**不应因对**咨询工程师（单位）**的索赔或声称的索赔，而扣留**协议书**规定的应适当支付给**咨询工程师（单位）**的任何款额发票的任何部分的付款，已与**咨询工程师（单位）**商定拟扣留款额应付给**客户**，或已由裁决员或仲裁员根据第 10 条［*争端和仲裁*］的规定裁定给**客户**的情况除外。

7.3 支付货币

7.3.1 适用于**协议书**的货币为**附录 3**［*报酬和付款*］中规定的货币。

7.3.2 如果在**协议书**生效日期或**服务**履行期间，工程所在国［**咨询工程师（单位）**的主要营业地除外］的条件可能如下：

（a）阻止或延迟将**本币**转到国外，或**咨询工程师（单位）**在**工程所在国**收到的**外币**付款；

（b）在**工程所在国**限制**外币**的可用性或使用；（或）

（c）对**咨询工程师（单位）**从国外将**外币**转入**工程所在国**作**本币**消费，以及随后将**外币**或**本币**再转至国外，征收相同金额的税款或差别汇率，如妨碍**咨询工程师（单位）**履行**服务**或对其造成资金上的不利影响。

则**客户**同意，如果替代资金安排未达到**咨询工程师（单位）**的满意，此类情况应被视为有理由适用第 4.6 款［*例外事件*］。[⊖]

7.4 第三方对咨询工程师（单位）的收费

7.4.1 除非**专用条件**或**附录 3**［*报酬和付款*］中有规定，并且除非**工程所在国**是**咨询工程师（单位）**的主要营业地，否则：

（a）**客户**应尽可能安排对**咨询工程师（单位）**及其通常不在**工程所在国**居住的人员，免除**工程所在国**政府或授权的第三方要求的与**协议书**有关的任何付款：

（i）其报酬；
（ii）其进口的食品和饮料以外的货物；
（iii）为**服务**而进口的货物；
（iv）为**服务**而进口的文件。

（b）如果**客户**未能安排此类豁免，则应为**咨询工程师（单位）**报销适当支付的此类款项；如果为**服务**而进口的货物或文件不再需要用于**服务**，而又非**客户**的财产，则：

（i）未经**客户**批准，不应在**工程所在国**处置；

（ii）在未向**客户**支付从政府或授权第三方收回和收到的任何退款或退费的情况下，不应出口。

⊖ 此处中文按照勘误表改正后的英文翻译。——译者注

7.5 Disputed Invoices

7.5.1 Without prejudice and subject to Sub-Clause 7.2.3 [*Time for Payment*], if any item or part of an item in an invoice submitted by the Consultant is contested by the Client as not properly due under the Agreement, the Client shall, within seven (7) days of the date of issue of the Consultant's invoice, give a Notice of its intention to withhold payment with reasons but shall not delay payment of the remainder of the invoice. Sub-Clause 7.2.2 [*Time for Payment*] shall apply to all contested amounts which are finally determined to have been payable to the Consultant.

7.6 Independent Audit

7.6.1 Except where the Agreement provides for lump sum payments the Consultant shall maintain up-to-date records which clearly identify relevant time and expense and shall make these available to the Client on reasonable request.

7.6.2 Except where the Agreement provides for lump sum payments, not later than one year after the completion or termination of the Services, the Client may, by Notice of not less than fourteen (14) days to the Consultant, require that an independent reputable firm of professionally qualified accountants nominated by it audit any time and expense records claimed by the Consultant. The audit shall be conducted by attending during normal working hours at the office where the records are kept and the Consultant shall afford all reasonable assistance to the auditors. Any such audit shall be at the Client's cost.

8 Liabilities

8.1 Liability for Breach

8.1.1 The Consultant shall be liable to the Client for any breach by the Consultant of any provision of the Agreement.

8.1.2 The Client shall be liable to the Consultant for any breach by the Client of any provision of the Agreement.

8.1.3 If either Party is liable to the other, damages shall be payable only on the following terms:

(a) such damages shall be limited to the amount of reasonably foreseeable loss and damage suffered as a direct result of such breach;

(b) in any event, the amount of such damages shall be limited to the amount stated in Sub-Clause 8.3.1 [*Limit of Liability*]; and

(c) if either Party is considered to be liable jointly with third parties to the other Party, the proportion of damages payable by that Party shall be limited to that proportion of liability which is attributable to its breach.

8.2 Duration of Liability

8.2.1 Notwithstanding any term or condition to the contrary in the Agreement or any legal requirement of the Country or any other relevant jurisdiction (including, for the avoidance of doubt, the jurisdiction of the place of establishment of the Consultant), neither the Client nor the Consultant shall be considered liable for any loss or damage resulting from any occurrence unless a claim is formally made on one Party by the other Party before the expiry of the relevant period stated in the Particular Conditions,

7.5			
有争议的发票	7.5.1		在不损害和遵守第 7.2.3 项［*付款时间*］的情况下，如果**客户**对**咨询工程师（单位）**提交的发票中的任何事项或部分事项提出质疑，认为其不符合**协议书**规定的应付款，**客户**应在**咨询工程师（单位）**发票开具之日起七（7）天内，发出拒绝付款的**通知**，并说明理由，但不得延迟支付发票的剩余部分。第 7.2.2 项［*付款时间*］应适用于最终确定应付给**咨询工程师（单位）**的所有有疑问的金额。
7.6			
独立审计	7.6.1		除**协议书**规定了一次性付款外，**咨询工程师（单位）**应保留最新记录，明确说明相关时间和费用，并应在**客户**合理要求时提供给**客户**。
	7.6.2		除**协议书**规定了一次性付款外，**客户**可在**服务**完成或终止后不迟于一年，通过向**咨询工程师（单位）**发出不少于十四（14）天的**通知**，要求由其指定的具有专业资格的会计师组成的独立而知名的公司，对**咨询工程师（单位）**申明的任何时间和费用记录进行审计。审计应在正常工作时间内到保存记录的办公室进行，**咨询工程师（单位）**应向审计员提供一切合理的协助。任何此类审计费用均应由**客户**承担。

8 责任

8.1			
违约责任	8.1.1		如果**咨询工程师（单位）**违反**协议书**任何规定，**咨询工程师（单位）**应对**客户**负责。
	8.1.2		如果**客户**违反**协议书**任何规定，**客户**应对**咨询工程师（单位）**负责。
	8.1.3		如果任一方对另一方负有责任，则只能按照以下条款支付损害赔偿费：
		（a）	此类损害赔偿费应限于因此类违约而直接遭受的可合理预见的损失和损害的金额；
		（b）	在任何情况下，此类损害赔偿费的金额应限于第 8.3.1 项［*责任限度*］中规定的金额；（以及）
		（c）	如果任一方被认为与第三方对另一方负有共同责任，则该方应支付的损害赔偿费比例应限于其违约造成的责任部分。
8.2			
责任期限	8.2.1		无论**协议书**中或**工程**所在国任何法律要求中或任何其他相关司法管辖区［为避免疑义，包括**咨询工程师（单位）**设立地的司法管辖区］的条款或条件是否有相反规定，除非另一方在**专用条件**中规定的相关期限到期前正式向一方提出索赔，否则**客户**和**咨询工程师（单位）**均不应对任何事件造成的任何损失或损害负责，该期限

such period to commence upon completion of the Services or termination of the Agreement (whichever is earlier). Each Party agrees to waive all claims against the other in so far as such claims are not formally made in accordance with this Sub-Clause 8.2.1.

8.3 Limit of Liability

8.3.1 The maximum amount of damages payable by either Party to the other in respect of any and all liability, including liability arising from negligence, under or in connection with the Agreement shall not exceed the amount stated in the Particular Conditions. This limit is without prejudice to any financing charges specified under Sub-Clause 7.2.2 [*Time for Payment*], and without prejudice to Sub-Clause 8.4.1. [*Exceptions*].

8.3.2 Each Party agrees to waive all claims against the other in so far as the aggregate of damages which might otherwise be payable exceeds the maximum amount payable under Sub-Clause 8.3.1.

8.3.3 Without prejudice to the right the Consultant may have under Sub-Clause 6.5.4 [*Effects of Termination*], neither Party shall be liable in contract, tort, under any law or in any statutory private right of action or otherwise, for any loss of revenue, loss of profit, loss of production, loss of contracts, loss of use, loss of business, third party punitive damages or loss of business opportunity or for any indirect, special or consequential loss or damage.

8.4 Exceptions

8.4.1 Sub-Clauses 8.1.3 [*Liability for Breach*], Clause 8.2 [*Duration of Liability*], and Clause 8.3 [*Limit of Liability*] shall not apply to claims arising out of deliberate manifest and reckless default, fraud, fraudulent misrepresentation or reckless misconduct by the defaulting Party.

9 Insurance

9.1 Insurances to be taken out by Consultant

9.1.1 The Consultant shall take out and maintain professional indemnity insurance and public liability insurance in amounts sufficient to cover its liabilities under the Agreement, provided always in each case that such insurance is available at commercially reasonable rates and on terms (including normal exclusions) commonly included in such insurances at the time the insurances were taken out or renewed as the case may be. Such insurances shall be placed with insurers of international repute and standing. In assessing a commercially reasonable rate the Consultant's own claims record shall be disregarded.
The Consultant shall ensure that the minimum amount of cover under the policies is not less than the amount specified in the Particular Conditions.
The Consultant shall ensure that its professional indemnity insurance is maintained for the period of liability stated in the Particular Conditions in accordance with Sub-Clause 8.2.1. [*Duration of Liability*].

9.1.2 The Consultant shall take out and maintain workers' Compensation insurance or employer's liability insurance and any other insurances as may be required by the applicable law for the duration of the Services.

9.1.3 When requested to do so by the Client, the Consultant shall produce brokers' or insurers' certificates to show that the insurance cover required by this Clause 9.1 is being maintained.

自**服务**完成或**协议书**终止时开始（以较早者为准）。每一方同意放弃对另一方的所有索赔，只要此类索赔不是根据第 8.2.1 项正式提出的。

8.3 责任限度	8.3.1	任一方就**协议书**规定的或与**协议书**有关的任何及所有责任，包括因疏忽而产生的责任，而须向另一方支付的损害赔偿费的最高数额不得超过**专用条件**中规定的金额。该限额不影响第 7.2.2 项 [*付款时间*] 规定的任何融资费用，也不影响第 8.4.1 项 [*例外情况*]。
	8.3.2	在可能应付的损害赔偿总额超过第 8.3.1 项规定的最高应付金额的情况下，每一方可同意放弃对另一方的所有索赔。
	8.3.3	在不损害**咨询工程师（单位）**可能享有的第 6.5.4 项 [*终止的影响*] 规定的权利的情况下，任一方均不对任何收入损失、利润损失、生产损失、合同损失、使用损失、业务损失承担合同、侵权行为、任何法律或任何法定私人诉讼权或其他形式的责任，第三方惩罚性损害赔偿费或商业机会损失或任何间接、特殊或后果性损失或损害。
8.4 例外情况	8.4.1	第 8.1.3 项 [*违约责任*]、第 8.2 款 [*责任期限*] 和第 8.3 款 [*责任限度*] 应不适用于因违约方故意和鲁莽的违约、欺诈、欺诈性失实陈述或鲁莽的不当行为引起的索赔。

9 保险

9.1 由咨询工程师（单位）承担的保险	9.1.1	**咨询工程师（单位）**应购买并维持职业责任保险和公共责任保险，保险金额应足以覆盖其**协议书**规定的责任，前提是，在每种情况下，此类保险以商业上合理的费率和在投保或续保时通常包含在此类保险中的条款提供（包括正常除外责任）。此类保险应向具有国际声誉和信誉的保险公司投保。在评估商业合理的费率时，应忽略**咨询工程师（单位）**自身的索赔记录。 **咨询工程师（单位）**应确保保单承保的最低保险金额不低于**专用条件**中规定的金额。 **咨询工程师（单位）**应确保根据第 8.2.1 项 [*责任期限*] 的规定，在**专用条件**规定的责任期限内维持其职业责任保险。
	9.1.2	**咨询工程师（单位）**应在**服务**期间购买并维持工人赔偿保险或雇主责任保险，以及适用法律要求的任何其他保险。
	9.1.3	当**客户**要求时，**咨询工程师（单位）**应出示经纪人或保险人的证明，以证明第 9.1 款要求的保险范围仍在维持。

9.1.4 The Consultant shall notify the Client immediately should any of the insurance required by this Clause 9 be cancelled by the insurers or underwriters.

10 Disputes and Arbitration

10.1 Amicable Dispute Resolution

10.1.1 If any dispute arises out of or in connection with the Agreement then senior representatives of the Parties with authority to settle the dispute shall, within twenty-eight (28) days of a written request from one Party to the other, meet in order to attempt to resolve the dispute amicably.

10.1.2 If the dispute is not resolved within fifty-six (56) days of receipt of the written request, then either Party may refer the dispute to adjudication in accordance with Clause 10.2 [*Adjudication*], even if the meeting referred to in Sub-Clause 10.1.1 has not taken place.

10.2 Adjudication

10.2.1 Unless settled amicably, any dispute arising out of or in connection with the Agreement may be referred by either Party to adjudication in accordance with the Rules for Adjudication in Appendix 5 [*Rules for Adjudication*]. The adjudicator shall be agreed between the Parties or failing agreement shall be appointed in accordance with the said Rules for Adjudication.

10.2.2 The Parties shall bear their own costs arising out of the adjudication and the adjudicator shall not be empowered to award costs to either Party. Without prejudice to the above, the adjudicator may decide which Party shall bear the adjudicator's fees and in what proportion.

10.2.3 If either Party is dissatisfied with the adjudicator's decision:

(a) the dissatisfied Party may give a notice of dissatisfaction to the other Party, with a copy to the adjudicator;
(b) this notice shall state that it is a "Notice of Dissatisfaction with the Adjudicator's Decision" and shall set out the matter in dispute and the reason(s) for dissatisfaction; and
(c) this notice shall be given within twenty-eight (28) days of receiving the adjudicator's decision.

10.2.4 If the adjudicator fails to give its decision within the period stated in the Rules for Adjudication, then either Party may, within twenty-eight (28) days of this period expiring, give a notice to the other Party in accordance with Sub-Clause 10.2.3, sub-paragraphs (a) and (b), above.

10.2.5 Except as stated in Clause 10.5 [*Failure to Comply with Adjudicator's Decision*], neither Party shall be entitled to commence arbitration of a dispute unless a notice in respect of that dispute has been given in accordance with Sub-Clause 10.2.3 or 10.2.4. If such a notice has been given, and neither Party commences arbitration of the dispute within one hundred and eighty-two (182) days of giving or receiving the notice, such notice shall be deemed to have lapsed and no longer be valid.

10.2.6 Whether a Notice of Dissatisfaction with the adjudicator's decision has been issued or not by either Party, any adjudicator's decision shall become binding on both Parties upon its release.

9.1.4　如果**第9条**要求的任何保险被保险人或承保人取消，**咨询工程师（单位）**应立即通知**客户**。

10 争端和仲裁

10.1 友好解决争端

10.1.1　如果产生因**协议书**或与**协议书**有关的任何争端，则有权解决争端的双方高级代表应在一方向另一方发出书面请求后二十八（28）天内举行会议，以尝试友好解决争端。

10.1.2　如果在收到书面请求后的五十六（56）天内未能解决争端，则任一方均可根据**第10.2款**[**裁决**]将争端提交裁决，即使**第10.1.1项**中提到的会议尚未召开。

10.2 裁决

10.2.1　除非友好解决，否则任一方均可根据**附录5**[**裁决规则**]中的**裁决规则**，将**协议书**引起的或与**协议书**有关的任何争端提交裁决。仲裁员应由双方商定，如未能达成商定，则应按照上述**裁决规则**指定。

10.2.2　双方应自行承担裁决产生的费用，裁决员应无权将费用裁定给任一方。在不影响上述规定的情况下，裁决员可决定由哪一方承担裁决员的费用以及按何种比例承担。

10.2.3　如果任一方对裁决员的决定不满意：

（a）不满意方可向另一方发出不满意通知，并抄送裁决员；

（b）该通知应说明其是"**对裁决员的决定不满意的通知**"，并应列出争端事项和不满意原因；（以及）

（c）该通知应在收到裁决员的决定后二十八（28）天内发出。

10.2.4　如果裁决员未能在**裁决规则**规定的期限内做出决定，则任一方可在该期限届满后二十八（28）天内，根据上述**第10.2.3项**（a）和（b）段的规定，向另一方发出通知。

10.2.5　除**第10.5款**[**未能遵守裁决员的决定**]所述的情况外，除非已根据**第10.2.3项**或**第10.2.4项**的规定发出有关该争端的通知，否则任一方均无权开始一项争端的仲裁。如果已发出此类通知，且任一方均未在发出或收到通知后一百八十二（182）天内开始对争端的仲裁，则该通知应视为已失效，并不再有效。

10.2.6　无论任一方是否发出了对裁决员决定的**不满意通知**，任何裁决员的决定一经签发，对双方均具有约束力。

	10.2.7	If the adjudicator has given its decision as to a matter in dispute to both Parties, and no notice under Sub-Clause 10.2.3 has been given by either Party within twenty-eight (28) days of receiving the adjudicator's decision, then the decision shall become final and binding on both Parties.
	10.2.8	Adjudication may be commenced before or after completion of the Services. The obligations of the Parties shall not be altered by reason of any adjudication being conducted during the progress of the Services.
10.3 **Amicable Settlement**	10.3.1	Where a notice has been given under Sub-Clause 10.2.3 [*Adjudication*] or 10.2.4 [*Adjudication*], both Parties shall attempt to settle the dispute amicably before the commencement of arbitration. However, unless both Parties agree otherwise, arbitration may be commenced on or after the twenty-eighth (28th) day after the day on which this notice was given, even if no attempt at amicable settlement has been made.
10.4 **Arbitration**	10.4.1	Unless settled amicably, subject to Clause 10.2 [*Adjudication*] and Clause 10.5 [*Failure to Comply with Adjudicator's Decision*], any dispute in respect of which the adjudicator's decision (if any) has not become final and binding shall be finally settled by international arbitration. Unless otherwise agreed by both Parties:

(a) the dispute shall be finally settled under the Rules of Arbitration of the International Chamber of Commerce;
(b) the dispute shall be settled by one or three arbitrators appointed in accordance with these Rules; and
(c) the arbitration shall be conducted in the ruling language defined in the Particular Conditions.

	10.4.2	The arbitrator(s) shall have full power to open up, review and revise any ruling or decision of the Adjudicator.
	10.4.3	In any award dealing with costs of the arbitration, the arbitrator(s) may take account of the extent (if any) to which a Party failed to cooperate with the other Party in appointing the adjudicator under Clause 10.2 [*Adjudication*].
	10.4.4	Neither Party shall be limited in the proceedings before the arbitrator(s) to the evidence or arguments previously put before the adjudicator to obtain its decision, or to the reasons for dissatisfaction given in the Party's notice under Sub-Clause 10.2 [*Adjudication*]. Any decision of the adjudicator shall be admissible in evidence in the arbitration.
	10.4.5	Arbitration may be commenced before or after completion of the Services. The obligations of the Parties shall not be altered by reason of any arbitration being conducted during the progress of the Services.
10.5 **Failure to Comply with Adjudicator's Decision**	10.5.1	In the event that a Party fails to comply with any decision of the adjudicator, whether binding or final and binding, then the other Party may, without prejudice to any other rights it may have, refer the failure itself directly to arbitration under Clause 10.4 [*Arbitration*] and Clause 10.1 [*Amicable Dispute Resolution*], Clause 10.2 [*Adjudication*] and Clause 10.3 [*Amicable Settlement*] shall not apply to this reference. The arbitral tribunal (constituted under Clause 10.4 [*Arbitration*]) shall have the power, by way of summary or other expedited procedure, to order, whether by an interim or provisional measure or an award (as may be appropriate under the applicable law or otherwise), the enforcement of that decision.

	10.2.7	如果裁决员已就争端事项向双方做出决定，但任一方在收到裁决员的决定后二十八（28）天内未根据第10.2.3项发出通知，则该决定应为最终决定，对双方均有约束力。
	10.2.8	裁决可在**服务**完成之前或之后开始。双方的义务不应因**服务**过程中进行的任何裁决而改变。.

10.3
友好解决

	10.3.1	如果已根据第10.2.3项［*裁决*］或第10.2.4项［*裁决*］的规定发出了通知，双方应在仲裁开始前努力友好解决争端。但是，除非双方另有商定，仲裁可在该通知发出之日后的第二十八（28）天或之后开始，即使未做出友好解决的努力。

10.4
仲裁

	10.4.1	除非已友好解决，否则，根据第10.2款［*裁决*］和第10.5款［*未能遵守裁决员的决定*］的规定，就裁决员的决定（如果有）尚未成为最终和具有约束力的争端，应通过国际仲裁最终解决。除非双方另有商定，否则：

（a） 争端应根据**国际商会仲裁规则**最终解决；

（b） 争端应由按照本**规则**指定的一名或三名仲裁员解决；（以及）

（c） 仲裁应以**专用条件**中规定的主导语言进行。

	10.4.2	仲裁员有权开启、审核并修改**裁决员**的任何裁决或决定。
	10.4.3	在涉及仲裁费用的任何裁决中，仲裁员可考虑一方未能根据第10.2款［*裁决*］的规定与另一方合作任命裁决员的程度（如果有）。
	10.4.4	任一方在仲裁员面前的诉讼中，均不应局限于先前为获得裁决员的决定而向其提出的证据或论据，或根据第10.2款［*裁决*］发出的一方通知中提出的不满意理由。仲裁员的任何决定应在仲裁中被接受为证据。
	10.4.5	仲裁可在**服务**完成之前或之后开始。双方的义务不应因**服务**过程中进行的任何仲裁而改变。

10.5
未能遵守裁决员的决定

	10.5.1	如果一方未能遵守裁决员的任何决定，无论是具有约束力的还是最终的和具有约束力的，则另一方可在不损害其可能拥有的任何其他权利的情况下，根据第10.4款［*仲裁*］和第10.1款［*友好解决争端*］的规定，直接将其提交仲裁，第10.2款［*裁决*］和第10.3款［*友好解决*］不适用于此项提交。仲裁庭（根据第10.4款［*仲裁*］组成）应有权通过简易程序或其他快速程序，命令执行该决定，不论是通过临时措施或暂行措施或裁决（根据适用法律或其他规定，视情况而定）。

10.5.2   In the case of a binding but not final decision of the adjudicator, such interim or provisional measure or award shall be subject to the express reservation that the rights of the Parties as to the merits of the dispute are reserved until they are resolved by an award.

10.5.3   Any interim or provisional measure or award enforcing a decision of the adjudicator which has not been complied with, whether such decision is binding or final and binding, may also order or award damages or other relief.

10.5.2 在裁决员做出有约束力但不是最终决定的情况下，此类临时或暂行措施或裁决应以明确保留为前提，即保留双方对争端案情的权利，直至通过裁决解决为止。

10.5.3 任何没有被执行的裁决员的决定的临时或暂行措施或裁决，无论该决定是具有约束力的还是最终和具有约束力的，也可以命令或裁决损害赔偿或其他救济。

ERRATA to the FIDIC Client/Consultant Model Services Agreement Fifth Edition 2017

The following significant errata **are not** included in the content of the Fifth Edition of the Client/Consultant Model Services Agreement. Several minor typographical errors and layout irregularities have also been found but are not included in this list due to their insignificance with regard to the content.

GENERAL CONDITIONS

Page 36	Sub-Clause 1.5.2:	On the fourth line, replace "services" with "Services".
		From the ninth to eleventh line, delete the sentence from "Either Party" to "where applicable" in its entirety.
Page 36	Sub-Clause 1.5.3:	Add a new Sub-Clause 1.5.3 as follows:
		"Either Party may require, by a separate Notice to the other, that the provisions of the Agreement be amended to comply with the change in legislation where applicable".
Page 44	Sub-Clause 2.1.2:	On the seventh line only, replace "Standard of Care" with "standard of care".
Page 46	Sub-Clause 2.4.2:	On the second line, replace "detailed particulars" with "supporting particulars".
Page 58	Sub-Clause 6.1.2(c):	At the end of the sentence, replace the full stop with a comma and add "subject to the Consultant giving seven (7) days' notice to the Client".
Page 60	Sub-Clause 6.3.3:	Delete the entire Sub-Clause 6.3.3 and replace it with the following:
		"6.3.3 If, during the suspension and resumption of Services or part thereof, the Consultant incurs Exceptional Costs, then:
		(a) as soon as reasonably practicable the Consultant shall inform the Client, by issue of a Notice, of the occurrence of these Exceptional Costs, and
		(b) the agreed remuneration shall be adjusted in accordance with Sub-Clause 7.1.2 [*Payment to the Consultant*]".
Page 60	Sub-Clause 6.3.4	Delete the entire Sub-Clause 6.3.4 and replace it with the following:
		"6.3.4 The Time for Completion shall be amended in accordance with Clause 4.4 [*Delays*] to reflect the effect of the suspension on the Programme.".
Page 60	Sub-Clause 6.3.5:	Add the following new sub-clause 6.3.5:
		"6.3.5 The Consultant shall take reasonable measures to mitigate the effects of the suspension of the Services or part thereof.".
Page 62	Sub-Clause 6.4.1(e):	On the first line, delete the phrase "Without prejudice to Sub-Clause 6.1.1 [*Suspension of Services*]," in its entirety.
		On the first line, replace "where" with "Where".
Page 66	Sub-Clause 7.3.2(c):	From the sixth to ninth line, delete all the words from "then the Client agrees that... ...to the satisfaction of the Consultant." and reintroduce the same wording as a new paragraph below, and applying to all of Sub-Clause 7.3.2.

PARTICULAR CONDITIONS PART A

Page 6	Sub-Clause 10.4.1	Delete the entire entry.

《客户 / 咨询工程师（单位）服务协议书范本》2017 年第 5 版勘误表[一]

第 5 版《客户 / 咨询工程师（单位）服务协议书范本》的内容中不包括以下重要修改。还发现了一些细微的排版错误和不规范版式，但由于其内容不重要，因此未列入本清单。

通用条件

第 36 页	第 1.5.2 项：	在第四行，以"Service"替换"service"。
		从第九行到第十一行，整句删除从"Either Party"到"where applicable"。
第 36 页	第 1.5.3 项：	增加下述第 1.5.3 项：
		"Either Party may require, by a separate Notice to the other, that the provisions of the Agreement be amended to comply with the change in legislation where applicable"。
第 44 页	第 2.1.2 项：	仅在第七行，以"standard of care"替换"Standard of Care"。
第 46 页	第 2.4.2 项：	在第二行，以"supporting particulars"替换"detailedparticulars"。
第 58 页	第 6.1.2（c）项：	在句末，以逗号替换句号，并加上"subject to the Consultant giving seven (7) days' notice to the Client"。
第 60 页	第 6.3.3 项：	删除整个第 6.3.3 项，并替换为以下内容：
		"6.3.3 If, during the suspension and resumption of Services or part thereof, the Consultant incurs Exceptional Costs, then:
		(a) as soon as reasonably practicable the Consultant shall inform the Client, by issue of a Notice, of the occurrence of these Exceptional Costs, and
		(b) the agreed remuneration shall be adjusted in accordance with Sub-Clause 7.1.2 [*Payment to the Consultant*]"。
第 60 页	第 6.3.4 项：	删除整个第 6.3.4 项，并替换为以下内容：
		"6.3.4 The Time for Completion shall be amended in accordance with Clause 4.4 [*Delays*] to reflect the effect of the suspension on the Programme"。
第 60 页	第 6.3.5 款：	增加下述新的第 6.3.5 项：
		"6.3.5 The Consultant shall take reasonable measures to mitigate the effects of the suspension of the Services or part thereof."。
第 62 页	第 6.4.1（e）项：	在第一行，全部删除第一行的语句"Without prejudice to Sub-Clause 6.1.1 [*Suspension of Services*],"。
		在第一行，以"Where"替换"where"。
第 66 页	第 7.3.2（c）项：	从第六行到第九行，删除"then the Client agrees that……to the satisfaction of the Consultant"中的所有词语，并重新引用同样措辞组成新的一段，并适用于整个第 7.3.2 项。

专用条件 A 部分

第 6 页	第 10.4.1 项	删除整个条目。

[一] 中文翻译文本已改正。本勘误表只是应 FIDIC 的要求而翻译，仅供参考。——译者注